Eyes Turned Skyward

50 years of the Ulster Aviation Society

1968—2018

ISBN 978-1-5272-3186-3

First published in 2018 by the Ulster Aviation Society

Designed by Jo-Ann Smyth, Evolution Graphic Design.

Printed in Northern Ireland by W&G Baird.

Funded by the Heritage Lottery Fund

Contents:

A Note from the Heritage Lottery Fund   4

Foreword   5

Introduction   8

Chapter 1:   Before the UAS   16
Chapter 1.1:   The Spotter's Diary   24
Chapter 2:   1968 - 1984   26
Chapter 2.1:   Ulster Airmail   44
Chapter 3:   1984 - 1997   52
Chapter 3.1:   How to Move an Aeroplane   92
Chapter 4:   1997 - 2005   102
Chapter 4.1:   Ground Bound Adventure   142
Chapter 5:   2005 - 2018   150
Chapter 5.1:   Hangar Tour   210
Chapter 6:   The Future of the UAS   220

Appendix 1: The UAS Heritage Collection   228

Appendix 2: UAS Timeline: Flightpath   230

Acknowledgements   235
Picture Credits   236

A Note from the Heritage Lottery Fund

Heritage Lottery Fund is proud and delighted to have ensured that National Lottery players have helped support the Ulster Aviation Society to collect, share and highlight 50 years of activity. A 50th birthday is a significant milestone for anyone and the Society is right to celebrate all they have achieved. They have gathered together an important collection of aviation technology relating to the island of Ireland and have lovingly restored, conserved and showcased the collection for others to learn from and enjoy. They bring the stories of incredible flying machines to life in a way which inspires, teaches and lets imaginations fly. We wish them every success as they write their next chapter.

Paul Mullan
Head of Heritage Lottery Fund NI

Stephen Riley is a retired television producer and journalist. He is the public relations officer of the Ulster Aviation Society and a past committee member.

He would like to dedicate this book to Fred Jennings for his friendship, humour, generosity and excellent example to all Society volunteers.

Stephen Hegarty is a design engineer with Bombardier Belfast and secretary of the Ulster Aviation Society.

He would like to dedicate this book to Uncle John, whose magazines and stories kindled a lifelong love of aviation, and whose gift of the National Air & Space Museum: An Autobiography *inspired this book.*

Air Vice-Marshal Harvey Smyth OBE DFC MA RAF
Patron, Ulster Aviation Society

Foreword

It is with great pleasure that I offer the foreword to this book, celebrating the 50th Anniversary of the Ulster Aviation Society. I am exceptionally proud to be Patron of the Society, and am constantly amazed by the commitment of its members, all volunteers who give of their time freely and selflessly to preserve and promote aviation heritage for Northern Ireland. With artefacts spanning from 'The Ferguson Flyer' to contemporary fast jets and helicopters, the Ulster Aviation Society supports our community far and wide, beyond the borders of the province, and has made the wonder of flight accessible to many who would otherwise not be presented with the opportunity. Hence, it is most apt that in its 50th year, the Society has been honoured with The Queen's Award for outstanding service as a voluntary organisation.

Man has always been intrigued by flight, and the aircraft that achieve it, as they 'slip the surly bonds of earth'[1] and afford their pilots the ability to literally reach for the skies: for many it is a passion which is unquenchable. I see this passion amongst the members of this Society, and whilst it is easy to focus on the aircraft housed proudly within the hangars at Long Kesh, it is the people of the Society that this book most importantly recognises, for without them, their unwavering dedication, and their relentless quest to commemorate the past and inspire for the future, the Society would merely be a collection of mechanical equipment. Looking to the future, it is heartening to see so many young people involved with the Society, be they Air Cadets or simply interested teenagers: against this backdrop, I am convinced that our next 50 years are secure, and will be as glorious as the last.

Air Vice-Marshal Harvey Smyth

1. *"High Flight"*, John Gillespie Magee Jr, 1941.

Aircraft occupy our attention, but people are at the core of the Ulster Aviation Society. Alan Moller is one of many volunteers, seen here working on our Fairey Gannet in 2017.

INTRODUCTION

In the spring of 2011, a short and somewhat chubby fellow, 75 years old and bearded, rattled open the door of the Ulster Aviation Society's hangar at the Maze/Long Kesh site and waddled into the cavernous space.

He gaped at the sight of one restored aircraft after another, spotlighted in the sunshine pouring through those high, wide windows.

He was Eddie Franklin, founder of the Society, which he had left with regret many years before to pursue another flying interest: bird-watching. With him was Hangar Boss Ray Burrows, who later recalled Eddie's visit: "Over the last few years he had bravely battled with cancer, a battle he finally lost…(He said) 'Jeez, I never thought the Society would come to this!' There was a sparkle in his eyes and pride in his voice. It was so obvious: Eddie had never lost his passion."

Perhaps he bequeathed that zeal to the tiny clutch of charter members who assembled the basic building blocks of the organisation 50 years ago. Their enthusiasm for flying machines has been channelled during that time by succeeding generations into a diverse range of activities and accomplishments which Eddie and his colleagues could only imagine back in 1968.

The membership of the Ulster Aviation Society, growing in size year by year, formalised their enthusiasm in 1985 with a constitution which would

It's no surprise, then, that the hobbies of aircraft spotting, photography and model building had taken hold in Northern Ireland.

govern their activities in the years ahead. But just as significant has been the continuing passion of the group, whose adaptability and hard work have seen it overcome momentous challenges. The Society has grown into a thriving hub of aviation education, restoration and preservation. And it was done entirely by volunteers—recognised eventually as a registered charity.

This book explores our 50 years of existence from the perspective of Society members past and present, garnered through interviews and research. Photographs have been selected from thousands of images to mark that progress. We've been assisted in the effort with a welcome grant from the Heritage Lottery Fund.

The intent is to provide a strong visual and documentary record of the Society, told largely from the perspective of the people who have made it the distinctive and successful enterprise which it is.

Our collection of 30-plus aircraft and displays is unique on the island of Ireland; nothing else of its size and variety exists here. But it's more than that. We guide thousands of visitors on group tours of

our two hangars, publish a monthly magazine and feature guest speakers at our monthly meetings. We also bring aircraft and exhibits to many community events each year, reaping an enthusiastic response from thousands more people. We arrange membership tours to other aviation collections and museums and provide guest speakers to other organisations. A select number of members have written books dealing with specific aspects of aviation history in Northern Ireland.

In the course of its lifetime, the Society itself has frequently made history, be it in the recovery of a ditched Wildcat fighter, the assembly of a replica Ferguson aeroplane or the strenuous and imaginative efforts to transport historic aircraft from England and Scotland to our two hangars at the Maze/Long Kesh site in Lisburn. That dynamism continues unabated. We believe that the efforts and energy which have brought the Society this far from our origins 50 years ago are reflected in this book.

The Ulster of 1968 was a hive of aviation activity. Civil traffic at Aldergrove airport increased as flying became accessible to all. On the military side, Royal Air Force Shackletons still

Every year thousands of visitors receive guided tours through the Ulster Aviation Society home at the Maze/Long Kesh, Lisburn. Visitors range from school groups to retirement clubs and everything in between.

Today the Ulster Aviation collection houses more than thirty aircraft, engines, uniforms, models, books, radios and other historic aviation artefacts as well as interpretative displays on aspects of aviation

A wonderful variety of aircraft could be seen in the skies at events throughout the island of Ireland. Air displays inspired budding aircraft spotters, many of whom are Ulster Aviation Society members today.

The Ulster Aviation Society has attracted significant media attention over the years, be it for the historic recovery of an aircraft or the political stumbling blocks which have restricted the group in recent times.

patrolled from Ballykelly and Army helicopters continued to serve in the province. No. 23 Maintenance Unit, formed during the war, was still active at Aldergrove, maintaining military aircraft. In the heart of Belfast, Short and Harland had just delivered the last of its mammoth Belfast transports, while production of its best-selling little brother, the SC7 Skyvan, was in full swing. New designs were being developed as well. Across the province, engineering firms employed local people through contracts to supply the aviation industry. It's no surprise, then, that the hobbies of aircraft spotting, photography and model building had taken hold in Northern Ireland.

There would be challenges, some of them not even anticipated, in organising and operating a society of aviation enthusiasts. For example, some newer members felt that aircraft spotting didn't quite spark their enthusiasm. They wanted to visit air shows, some of them a good distance away. Others were interested in history and its physical remnants in the shape of aircraft wreckage scattered about the province. And right from the start, the spectre of impending conflict in the province hovered over the organisers. There were consequences, but the little Society struggled on.

The members met regularly over the years—bar occasional interruptions during the years of the Troubles. They also organised trips, raised some money and lost some. They adapted to meet other aviation interests. They acquired some aircraft bits, bounced from one meeting site to another, and faced eviction but rallied with resilience. Political disagreement in recent years has dogged the Society's efforts, but they have prevailed in the face of change and obstruction. They don't court controversy, but they don't shy away when their interests appear threatened. Some of their best people have come and gone in the past half century, but still the membership has grown, from less than a dozen in 1968 to over 500 by now. Their magazine, *Ulster Airmail*, has been a major thread in holding them together, a component in the determination which has remained central to their longevity. Couple that resolve with the enthusiasm and hard work of their volunteers and you have the ingredients for a unique community which has not just survived, but thrived for 50 years.

We hope you enjoy exploring the story of the people who have built the Ulster Aviation Society from the ground up to become Northern Ireland's foremost custodian of aviation history. It's been a struggle at times, but the passion remains.

Volunteers like Harry Munn (right) have given up their time to bring items from the collection to the public at about 20 events per year. Here, a lad grabs the chance to sit inside the cockpit of a replica Spitfire.

Exhibits in the Ulster Aviation collection are restored and maintained by a passionate community of volunteers–people like Leonard Craig, seen here. Others guide visitors, clean exhibits, manage paperwork, record restoration progress, research aviation history, give talks and plan for the future.

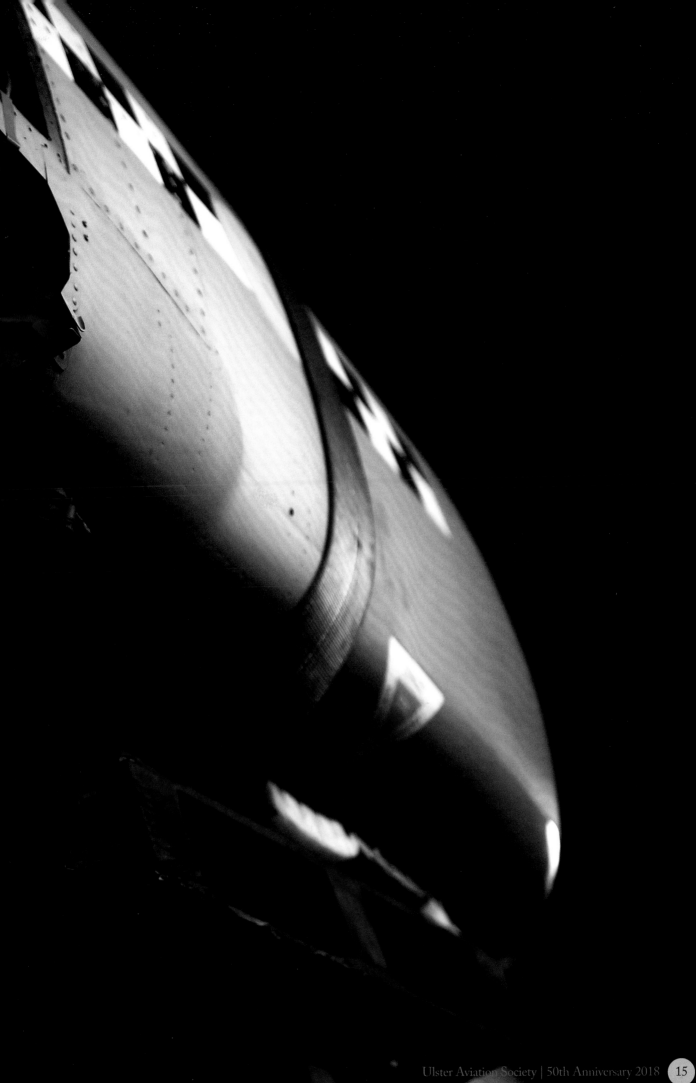

CHAPTER ONE

Before the birth of the Ulster Aviation Society, its foundation was gradually forming–unaware even to the people behind it. The enthusiasts had known in their innocent youth that aviation was special.

They might not say so, but powered flight seemed semi-magic, a fulfilment of dreams which had intrigued humanity for eons. And the machines of the post-war era which made it so captivating seemed to be the creations of wizards. The founding members of the Society recognised the same appreciation in others; they somehow knew they were kindred spirits, partners in an odd elite. They spoke the same arcane language—the argot of aviation. Of course, over time the uninitiated sometimes used their own language, labelling the aviation buffs as nerds, geeks and the like.

> "I don't know what my neighbours thought of me. Probably thought I was absolutely nuts!"

Ray Burrows, an enthusiast since the 1950s, has leapt to the defence: "It wasn't a childish pursuit. It was a serious thing. I lived in east Belfast at the time, on the base leg for 04 runway for the City Airport, and when I heard something I raced out, jumped on the bin and climbed up on the top of the yard wall with my telescope. I don't know what my neighbours thought of me. Probably thought I was absolutely nuts!"

In their teenage years, the spotters' schoolmates had babbled half-knowingly about automobiles: the speed and styling of the Jaguar E-type and semi-exotic MGB or the saucy design of the blockish and zippy Austin Mini. But most young aviation enthusiasts set their sights higher, into the blue. They would thrill at the sight, or even the thought of a Gloster Javelin or an English Electric Lightning slicing the sky at many times the speed of those earthbound dawdlers. And their drivers were genuine heroes, actually tickling disaster from take-off to landing. In their daydreams the young fans revered them— perhaps even aspired to one day become heroes themselves.

The Society's blossoming founders and followers fed one another those fantasies. They loitered at magazine stands, flipping through *Air Pictorial* or *RAF Flying Review,* often without the pocket money to afford every monthly issue. The stories and pictures within told of daring new designs soaring from the drafting tables: Concordes, Phantoms, Mirages, MiGs and the amazing TSR-2. With a bit of extra saving, they could head for the picture houses. As children, they might have seen The Sound Barrier or The Dam Busters, and by the late 1960s there was talk of a new and challenging film in production: The Battle of Britain. How terrific that would be! Inspired, they scrambled home to build their Airfix Spitfires, Hurricanes and dastardly Messerschmitts —lads in their aircraft habitats. And it was a boys' pastime, that much was obvious. Some diehard builders are still at it 50 years later, with improved materials and tools, their plastic models fashioned to a quality never imagined. But they, like most enthusiasts of the 1960s, had bigger aircraft in their sights.

Northern Ireland has had an

The Mosquitoes, the aerial action and that amazing theme music! They made 633 Squadron one of many exciting aviation films from the postwar period.

The speed, the deafening roar and the unique design of the English Electric Lightning thrilled air show audience. Young, aspiring pilots dreamed of flying this fighter. Above is a P1B prototype.

aviation heritage out of all proportion to its population. There were early heroes, like Harry Ferguson, Lilian Bland and Joseph Cordner, plus a core of flying veterans from the two world wars. They brought with them stories of high flight and future possibilities. Each day saw passenger aircraft arriving from a worldwide airline industry. It was rapidly expanding, with new, jet-powered airliners. They were fast, safe, comfortable and capable of ever-increasing ranges. Air fares were decreasing, to the point where average workers could eventually afford flights to faraway places which up to then they had only been reading about. By 1968

Northern Ireland had a sizeable airport at Aldergrove, a smaller operation at Sydenham and sporadic service at Eglinton. There was a presence as well of military flying units, albeit considerably smaller than it had been during wartime. Two of the service airfields, at Langford Lodge and Long Kesh, would have future importance for the Ulster Aviation Society. Flying machines of all sorts were coming and going in Northern Ireland. Helicopters, once a rare sight, were becoming more common in the skies throughout the province.

Meanwhile, the group of post-war enthusiasts were growing up. Within that mixture, a

Two key magazines for enthusiasts in the 1960s and '70s. **Flying Review** *had the exciting stories; the more reserved* **Air Pictorial** *had more technical details and a bit of aircraft registration data.*

specialised but unorganised group tended to gather wherever aeroplanes and helicopters could be seen at relatively close quarters. These were the aircraft spotters or (less laboured in American slang), "plane spotters." And a spotter was not just a sky-watcher: You might be a millo or a civvy. Or fascinated by a reggie or a dotty. In the spotters' lexicon (and not all of them were, or are, familiar with the terminology), the terms denoted a dominant interest in either military or civil aircraft, or in their registrations. A dotty was someone with higher aims who tried to discern types or markings on high-altitude aircraft—say, at 20,000 feet or more, mere dots in the sky. In any case, plane spotting was often deemed by non-enthusiasts to be a rather odd interest.

"But we didn't think of it like that," said David Hill of Londonderry, whose aviation enthusiasm began as a teenager in the early 1960s. "We were just keen spotters. I used to cycle down to the naval air station at Eglinton every Saturday… and the aircraft were there: Gannets, Fireflies, Avengers! And you could walk to the guard gate, and it was no bother. They just let you stand there and watch the aircraft taxi past." His passion has been lifelong, merging over the years with model-building of the highest calibre, winning him prizes regularly in competition. The spotters' preoccupation comes to them honestly, and with a historic pedigree. Careful observation, after all, goes back to primitive humanity: It was important to recognise potential danger at a distance, in the form of a large carnivore, a poisonous serpent or maybe a stranger with a spear. Or, equally important, a possible ally.

In more modern times, such caution has taken on new forms. The spotters of the Royal Observer Corps in the Second World War bore an equal responsibility for the survival of their national tribe, a duty that new-fangled radar never fully replaced at that time. The ROC had no bases in Northern Ireland at the time, but their reputation was nation-wide. When not gazing skyward and reporting aircraft movements, they could study their sheets and posters of aeroplane silhouettes. (Several of those images, in three-dimensional form, ended up as accident wrecks in the

The expression on this mannequin, in Royal Observer Corps uniform, indicates serious business. She and the postwar poster are displayed in our Royal Observer Corps Rooms. They're two of many items on exhibit, thanks largely to the tender care proffered by UAS member Paddy Malone.

UAS member David Hill has been a dedicated spotter for over half a century in Ulster's northwest as well as a regular prize-winner at modelling competitions. These models, of two Shackletons once spotted at RAF Ballykelly, won David the Society's annual modelling competition in 2016.

Spotters on the job: Belfast City Airport remains a popular site for spotters, though the variety of aeroplanes has diminished considerably during the past 50 years.

hills, meadows and bogs all over Ireland. Some of the ROC people, and the local watchers who emulated them, doubtless made mental notes of crash locations, for possible future reference.) Many of the more observant spotters could become intrigued, fixated even, with aircraft details. And in the post-war years, that fascination spread beyond the ROC, becoming a hobby to small clutches of observers who gathered at the edges of airports.

Obviously, the closer a spotter could get to an aircraft, the better to see such details. And how close was close? Certainly not so near that the airport police might appear on the

scene. In fact, Eddie Franklin, a keen spotter through the 1960s, warned his fellow enthusiasts in a comment column to respect the ground staff at Aldergrove: "We have built up a happy system of co-operation, especially with the police," he noted in *Ulster Airmail*. "We can maintain this understanding by using our common sense...."

Firm words and fine. But fellow enthusiast John Barnett has recalled a fairly lax system at the time—one of which he and Franklin occasionally took full advantage: "In those days you could've done anything around the airport.... Nobody was looking for security passes or anything. It was just free and easy."

> "The aluminium crew ladder was down, and quite illegally we both clambered up the ladder and had a look at the flight deck," Barnett recalled many years later.

Just how easy is typified by a day at Aldergrove when the pair learned of the imminent arrival in 1970 of a DC-7C passenger aircraft with Bermuda registration. An unusual visitor, just the kind to quicken a spotter's pulse. But Barnett and Franklin were poorly placed to photograph the aeroplane's approach. Frustrated at missing the shot, they sped around to the arrival apron at the terminal and took their photos from there. Passengers and crew departed the aircraft. The cleaning team did its job and headed into the terminal. Franklin and Barnett edged a bit closer to their quarry and spotted their chance.

"The aluminium crew ladder was down, and quite illegally we both clambered up the ladder and had a look at the flight deck," Barnett recalled many years later. "We spent maybe five minutes inside, just looking around. It was quite remarkable!"

Evidence of a trespasser: Eddie Franklin is caught on John Barnett's camera descending from the DC-7C. Even the most dedicated spotter wouldn't try such a stunt these days.

Anyone attempting a stunt like that these days can, of course, expect a well-armed cordon of police, weapons drawn, as a welcome party. As it was, Barnett managed a quick photo of Franklin's rump as he descended to the ramp– clear evidence of misplaced enthusiasm.

The spotters tended to find easier ways to satisfy their close-up interests, often by attending air shows, most commonly during the 1960s at the Aldergrove, Sydenham and Newtownards airfields, but also at Dublin or in Scotland, England and even beyond. At such venues, they discovered others with the same interests. And it wasn't just the air shows. Dublin was a favourite destination for many Northern spotters in the 1960s, especially if rugby fans among them could combine a bit of spotting with a trip to catch a Five Nations Championship. Such trips provided opportunities as well to exchange notes and photographs with aviation aficionados in the Republic.

Posters like this were a vital part of an observer's resources during the Second World War. Defence publications updated the information on a regular basis.

This Douglas DC-7C, registered in Bermuda, was an exceptional arrival at Aldergrove. It was 1970, when jet aircraft were the common visitors, so Eddie Franklin and John Barnett dashed over to Arrivals Gate 7 for a closer look at a relic of aviation history.

John Barnett was an enthusiastic spotter and photographer well before he became a charter member of the Ulster Aviation Society. Here he was in November, 1969, galloping closer to his subject for a better shot.

There was an Irish Society of Aviation Enthusiasts in the 1960s. Within it was a loose Ulster branch, led by Eddie Franklin. These Northern spotters had been considering a breakaway from the main organisation.

There appears to be no official record for the reasons behind their discussions. However, it can be safely assumed that the driving time between Belfast and Dublin could be discouraging. Congested road traffic through small communities like Newry, Dundalk, Drogheda and Balbriggan did nothing to help matters. And, simply put, spotters from the Republic weren't regular visitors to airports further north. Northern

Ireland's spotting community may have felt at times like poor, isolated cousins. True, airports at Dublin and Shannon offered much for those with interest in commercial movements. On the other hand, Northern spotters with military interests were blessed with a variety of Royal Air Force and Royal Navy movements.

Up to the end of the 1960s, the idea of separation was just talk. But on December 15th, 1968, when the Northern spotters met as usual at Aldergrove's old Gate 7, all that changed dramatically. Franklin laid it out simply: A body should be formed solely for Northern Ireland enthusiasts. The response from his colleagues was unanimous. Three weeks

Franklin laid it out simply: A body should be formed solely for Northern Ireland enthusiasts.

later, he advised the annual meeting in Dublin of the Irish association that Ulster spotters were striking out on their own. They had even adopted a name: The Ulster Aviation Society.

Gradually, in the years to come, the little group of spotters would grow, catering to other enthusiasts with diversified interests. There would be modellers, radio buffs, pilots, historians, photographers and many members who just liked hanging around aircraft.

They were practical people, but among them were a few who also imagined what might be possible. The dreams had begun; the challenge now was to build on them.

The year is 1968, the airport is Aldergrove and a DC-3 has arrived, to the delight of spotters. It was a symbol of a diminishing, prop-driven era.

Aircraft spotters were among a group of our Society members who got a close look at this rare aircraft on a trip to Dublin Airport in the late 1980s. The visitor was a Spanish-built Ju52/3m. This particular aeroplane is now an exhibit at a Swedish aviation museum.

This is where it all began, at Belfast International Airport, Arrivals Gate 7, its white-framed observation deck on the first floor (to the right of centre). A little cluster of aircraft spotters gathered here on December 15th, 1968 and cheerfully endorsed Eddie Franklin's suggestion to form an organisation of enthusiasts. They would call it the Ulster Aviation Society.

The Way It Was

Ernie Cromie:

I became interested in aviation when I was five years old. That was in the very bad winter of 1947 in County Down, and the snow was so deep there in the foothills of the Dromara mountains. Just like a vast, white desert. One day, I heard the sound of aircraft engines. I looked up and there was this large, black, four-engined aircraft circling and fodder being thrown out of it for livestock. And that was the first time I had consciously seen an aircraft. I was just blown away by the experience, and ever after that I developed and maintained an interest in aviation-particularly the historical aspect, but spotting as well.

John Barnett:

It was a remarkable thing as an aircraft spotter to watch people walking out towards a Viscount, watching it taking off and knowing that in 40 minutes they'd be touching down in Liverpool, or in London an hour and ten minutes later. And you'd be thinking, 'Goodness! I wish I could be there!' It was just a remarkable thing that the speed of travel was so fast and so romantic.

Ray Burrows:

On a good day, there was nothing as nice as looking at a blue sky and seeing those beautiful silver aircraft with red markings or red and white markings flying past, absolutely glistening in the sky. And I always wondered, you know, 'Who's on board that? Where are they coming from? Where are they going to?' Totally fascinating!
The diehard spotters? Well, I always considered them to be nuts! I always considered myself to have a healthy interest in the subject but I saw these guys with maybe five thousand different serial numbers in their little book and ticking each one off and the date they saw it, and I thought 'That's a bit screwy! That's going a bit too far.' But I was a bit like that, I suppose, in my teens.

1.1 | The Spotter's Diary

In piecing together this Society history, one issue cropped up again and again: Where are all the photos from the first ten years?

Most spotters, it seems, were interested in simply viewing aircraft and recording their registrations and movements. A few were into photography, but most were amateurs in that respect, and the results frequently showed it, with subjects out of focus or far in the distance. Their equipment was simple—often just a Kodak Brownie or Instamatic film camera, relatively cheap but limited in their capabilities. The photos themselves were usually black and white. Colour photography was expensive for many of our members and the results were often lacking in high quality. Detail, contrast and colour rendition left much to be desired.

And what exactly did spotters do? There are few, if any photographs showing our earliest members in action. As Jack Woods—himself a serious photographer—put it, do you think the members were going to point their cameras at each other or at the planes?

Over time, techniques and equipment have undergone a revolution, with digital imagery replacing analogue film, and single-lens reflex cameras which incorporate complicated options. Aircraft photography in the Society has evolved from casual interest to a serious hobby, with results produced at a professional level. Each December, the intrepid photographers compete for prizes, with their best efforts on show. In the days and months in between, their work appears online at our website and on Facebook, and each month Ulster Airmail carries a selection of the latest spottings, produced in fine colour and detail.

Here are a just a few aircraft shots our members have snapped over the years to share with their peers, or perhaps enter into a competition, but always to record a single moment in time when the unrelenting thrill of a flying machine passing overhead inspired them to turn their lenses skyward.

Founder Eddie Franklin was himself a keen photographer and captured this beautiful image of a mighty Shorts Belfast taking off.

UAS spotters are often on hand to capture national and international 'firsts,' in this case EI-ASA: Aer Lingus' first 737.

This 1963 arrival of de Havilland Heron G-AZON at Aldergrove predates the UAS. The formation of the Society turned individual photographers into a community.

The first commercial blimp to visit Northern Ireland was an easy target for the UAS photographers!

Society trips allowed spotters to add unusual aircraft to their logs like this Super Guppy seen at Manchester in 1984.

Jack Woods shot this rare Lockheed Jetstar at a Shannon air display in the 1980s.

February 7th 1983: First to touch down at the re-opened Belfast Harbour Airport was this Spacegrand Twin Otter.

A lone spotter snaps a Fokker F50 on approach at Dublin.

From its earliest days the Society magazine has provided a platform for photographers to share their spottings.

CHAPTER TWO

Flying an aeroplane for the first time is daunting enough, beginning with flight preparations and take-off. Preparing the Ulster Aviation Society for flight proved to be no simpler during its earliest days.

As one of the early members put it years later: "If we'd guessed then at the kinds of challenges we'd be facing, I doubt we ever would have gone ahead with it."

The birth of the new group threatened at first to become less of a celebratory launch than akin to a snarly divorce. Some three weeks after formation, a trio of new Ulster Aviation Society members, including Eddie Franklin, headed for Dublin. The annual meeting of the Irish Society of Aviation Enthusiasts was set to convene. The northern trio was about to announce a separation. The date was January 4th, 1969, and partition was apparently not on the agenda.

"(T)he meeting tended to get a little hot at times," according to an understated report in the new UAS journal, calling itself *Ulster Airmail*. The details of those boiling points were not revealed, but at least by the end of the gathering the northern representatives "left in a friendly atmosphere with the promise of full co-operation between both groups."

That north-south link has proven to be sporadic through the years, but there was no

The founding chairman gathered around him a core group which would aid in the work at hand.

doubt that the hopes for a positive relationship were genuine.

For his part, Eddie Franklin returned with no intention of letting his new flock relax and preen its proud feathers. There was work to be done and the first issue of *Ulster Airmail* in February, 1969 was evidence of that.

It might seem a curiosity that so small a group—no more than two dozen in the earliest days—would see a need for a regular publication, except for one thing: They wanted a forum in which to list, month after month, the reams of aircraft movements and registrations, the very lifeblood of spotters. That first issue alone was only eight pages in size but more than half of them were packed with lists of aircraft movements and registrations.

The eyes of less interested enthusiasts might glaze over, but for those unfamiliar with the form, here is one small entry in *Ulster Airmail* from the Aldergrove spotters' log for a few days in November, 1968: *"DI-EVW King-Air Volkswagen; G-AVFG Trident 2 (morning) and G-AVFD (evening). 13th – G-AVFC Trident 2; G-ATZJ*

Aztec 250 Unigate Creameries. 17th G-AWPP Cessna F150H J. Dalgleish."

So, five pages of spotter language. And if an enthusiast yearned for data from more exotic climes, the magazine advised readers that member Chris Mahaffy was there to help: "He has over 90 registers covering such odd places as Upper Volta, Dahomey and some that we can't spell."

In any case, you had to be committed (and some joked about that description), but it bears reminding: Spotters were the bedrock of the Society and their record of aircraft comings and goings continues to be included in the magazine, month after month every year. At the very least it's a tribute to their dedication. As well, perhaps, it's recognition from the Society at large of the spotters' contribution during the past half century.

By all accounts, that small coterie of original members was a buoyant bunch, and none more so than Chairman Franklin. He was a keen plane spotter, a short, rotund Scouser who enjoyed his pint, and never more so than during those first meetings, which had shifted from Gate 7 at Aldergrove to the First and Last pub in his adopted home town of Antrim.

"Eddie was a jovial, very welcoming guy," recalled Ray Burrows. "Just one of these guys who made you feel at home from the moment you met him, and we got on like a house on fire."

Throughout the Society's early years, its meeting venues changed with a peripatetic regularity. Within weeks, the membership had moved their gatherings to nearby Hall's Hotel, where the format changed from relaxed nattering and picture exchanges to something only slightly more formal.

"We had a ten-minute talk, then watched people's slides," said Ray. "It was just a general sit-down over a pint of beer and a chat about what we were going to do next month." And the chats often continued late into the night.

"They'd come to our house for coffee afterward," recalled Eddie's son, Trevor. "They'd meet in the front room, but we had to stay in the back room, and Mom would make them sandwiches."

The founding chairman gathered around him a core group which would aid in the work at hand. Among the members were Maurice Hatrick, Roger Andrews, Hugh McGrattan, Chris Mahaffy, David Adgey, Lowry McComb, Con Law, Trevor and Wesley Haslett, Ross

MacKenzie and David Hill.

The latter two enthusiasts came from the Derry/Londonderry area, and brought information to the meetings about aircraft types and movements–very welcome details for the other members, gorged as they were on Belfast and southern airports data.

"We were in a backwater in the northwest," said Ross. For commercial aircraft, certainly, but military aircraft visited RAF Ballykelly regularly, he added: "So it was a very vibrant aviation scene up there... A lot of NATO aircraft visiting from Canada, Norway, the United States, Holland, France, Portugal ... all coming for two or three weeks at a time."

From that area as well came early suggestions by a short-lived subgroup that aircraft wrecks around Ireland could be located and recovered. Aircraft restoration and preservation were clearly becoming matters of increasing historical and active interest, not least in the *Ulster Airmail*. It was publishing regular, well researched stories on our aviation heritage, including a few by a new member, one Raymond Burrows. In 1973, the magazine published a Burrows article on wreckology—a foreshadowing of what might eventually happen with the Society. The magazine had also noted a suggestion as early at 1969 that Aer Lingus planned to set up an aviation museum at Dublin Airport. Minor efforts were discussed and even launche

Eddie Franklin, 1968. His enthusiasm, determination and good humour gave him a firm grounding to lead the Ulster Aviation Society in its formative years.

> "It was just a general sit-down over a pint of beer and a chat about what we were going to do next month."

at times, but little came of them over the years. To date, the only significant museum in the Republic is run by the Irish Air Corps at Baldonnel, with 12 complete airframes and some memorabilia but limited access.

The Ulster Aviation Society, for its part, was forging ahead in a cautious but determined manner. There were photo/modelling competitions, calendars and a bit of low-level fund-raising as well as continued meetings and p u b l i c a t i o n — a l b e i t occasionally sporadic in the 1970s—of *Ulster Airmail*.

Monitor the radio airband, spot the aircraft, maybe shoot a photo, but definitely record the registration. If you could do that, perhaps you and your mates could form an enthusiasts' society.

©Victor Patterson

Troops man a vehicle checkpoint near Aldergrove Airport in the 1970s. Regular patrols and warning signs discouraged spotters from loitering anywhere nearby.

Hall's Hotel, in the aftermath of a bomb in 1972. It was one of the early meeting places of the Society. Members moved to another venue just weeks before this incident, one of hundreds that plagued the province that year.

Membership was slowly growing (to about 40 by 1973) and, though many enthusiasts were consumed with interest in aviation matters, that interest was increasingly compromised by events outside their control.

Aircraft hijacking, especially by radical groups, became an international epidemic in the 1970s. On the America-Cuba routes alone, at least 30 acts of air piracy took place in 1970, the worst year ever. Security measures around airports worldwide became tighter and with the conflict here in Northern Ireland at a very serious level, plane spotters and photographers were finding their hobbies increasingly restricted. "Security was a big, big issue," said Ray Burrows. "You couldn't go to Aldergrove and park your car and get out with your camera and take photographs of aircraft. You would have been moved on,

> Plane spotters and photographers were finding their hobbies increasingly restricted.

possibly arrested, within five or ten minutes."

The airport itself was targeted by the Irish Republican Army on March 6th, 1976, with 13 mortar bombs hitting the site. Fortunately, there were no casualties and only minimal damage. Airport security forces were discredited by such a blatant gesture, however, and stepped up their measures limiting access to areas near the airport.

Shackleton anti-submarine aircraft were a common sight in the northwest while they were based at RAF Ballykelly during the 1950s and '60s. UAS spotters watched the last one leave in 1971.

Admittedly, military flights of troops and equipment provided limited spotting opportunities for those who could manage to avoid security precautions, but timetables were non-existent, so making the effort seemed pointless at times. And where was the joy in confronting uniforms on a regular basis?

Eddie Franklin was one of several who struggled to maintain a passion for spotting and photography through the Troubles. However, the difficulties imposed by the

Eddie Franklin was one of several who struggled.

new security measures vastly outweighed the reward of spotting a rare craft on the approach.

He decided in 1970 to limit his own Society activity by resigning as chairman. Hugh McGrattan stepped up as "honorary chairman" in 1972 to fill his place. Eddie remained active at a lower level, but his continued involvement seemed to have a limited duration.

The Society's magazine archive, along with older members' memories, testify to occasional doubts through the 1970s that the organisation could even continue. Society meetings—which had bounced from one Antrim location to another—were sporadic in 1976, and the *Ulster Airmail's* editor had to regularly plead for articles.

"That the Society survived at all is nothing short of miraculous," Ray Burrows reflected a few years later. However, the core membership remained determined and some of the concerns about limited local opportunities were allayed by a simple solution: Look elsewhere to improve the situation. Members took trips to Greenham Common, Leuchars, Prestwick and Dublin for air shows. Those ventures seem to have re-ignited interest. In fact, such was the demand to reinstate the monthly gatherings that in July 1977 the *Airmail* carried a rare upbeat notice that meetings were to be reconvened in the coming autumn.

Momentum gathered and in September 1977 the Society issued the largest magazine in its nine-year history: 16 pages! That same month, Eddie Franklin and Roger Andrews had the pleasure of being

Early UAS member Ray Burrows was one of a small UAS group which visited the Lough Foyle Corsair in 1973. Recovery of the remains from the large stretch of tidal ooze was deemed impossible.

chauffeured to the Battle of Britain air show at RAF Leuchars by fellow member and sixteen-year-old pilot, Con Law. Con's employers at Woodgate Aviation were happy to let the young trainee borrow Cherokee G-AZVV for the day, under supervision from Shorts test pilot John Richardson. This wasn't a unique occurrence—Con had previously flown UAS members on a day-trip to Shannon Airport.

Finally, on October 18th, 1977 the membership gathered for the first time in almost two years at the new location of Antrim Technical College, where Roger Andrews shared his collection of slides from that year's Greenham Common air display. Ray Burrows remembers the new

"That the Society survived at all is nothing short of miraculous"

meeting venue fondly: "the classroom next to us was domestic science ... we were being wafted with scents of apple tarts. We all went home rather hungry!"

After years in limbo the group had finally taken its first tentative step back towards some semblance of normality. But as the Society's 10th anniversary approached, a magazine column from Eddie Franklin put the situation into perspective: "How many people who attended the first meeting at Aldergrove on December 15th, 1968 honestly thought we would last ten years?...I for one had very serious doubts, just as I now do about the next ten years."

One of the most obvious wrecks in Northern Ireland has been this *Corsair II (JT693)*, which crashed on the Lough Foyle mudflats after takeoff on October 9th, 1944.

*Here's one we didn't see at the time. Restricted access at RAF
Aldergrove meant no enthusiasts' cameras on site. But an Army
helicopter crewman snapped this embarrassing shot in 1977 of a
Phantom FG1 (XV572) mired in mud after veering from the runway.*

*Woodgate Cherokee G-ATWO. Her sister aircraft G-AZVV ferried
Eddie Franklin and Roger Andrews to the 1977 Battle of Britain
Air Show at the hands of young member Con Law, all glad to
escape the security restrictions which hindered local spotting.*

*UAS members enjoy each other's company and the closeup sight of an Aer Turas CL-44 on the ramp at Dublin Airport. In the 1970s and
'80s, visits to airports outside the North were a delightful option for enthusiasts. (Aer Turas, a cargo business, operated out of Dublin from
1962 to 2003.)*

Ray Burrows was troubled by Eddie's agitation about the Society's future: "He expressed his opinion that things weren't going well, and Eddie was the anchor man of the Society for many years. I used to think that if Eddie's feeling down then the whole Society was feeling down."

Certainly the *Ulster Airmail's* little production team would have agreed. The magazine was the glue binding the Society together. It didn't always reflect the membership's mood in general but as the year rolled by the pleas for stories became more serious. "Enough is enough," it warned in one edition. And it went further: "If things do not improve shortly then our Society will come to an end at the end of this year."

The history of the Ulster Aviation Society could well have finished there. It could have been the solemn obituary of keen hobbyists, united by a love for all things aviation but ultimately overcome by frustration as the airports of their youthful spotters' days closed their doors against them.

Theirs is not a story of defeat, however. It is one of fervent and dedicated volunteers who, time after time, rallied to the rescue of their clan. Such was the case in October of 1979 when no fewer than nine volunteers signed up to help with *Airmail* production.

> "Eddie was the anchor man of the Society for many years...if Eddie's feeling down then the whole Society was feeling down."

Perhaps the cries of desperation had done the trick. Perhaps there was an underlying feeling that the Society could be more than it seemed. Whatever the stimulus, there were reasons for cautious optimism. Membership in 1980 was rapidly approaching 100 and the annual fee was reasonable—only £2.50 a year. And there was money in the bank, a grand total of £20. Most important, the organisation had survived yet another difficult year. And it turned out that despite the problems encountered in the 1970s, the members had

The Antrim Technical College was another early meeting place for the Society. Antrim was convenient to spotters, being close to Belfast International Airport and the RAF base. Chairman Eddie Franklin, who lived in Antrim, found the college venue handy as well! Members who lived in Belfast found night-time driving to the Antrim meetings less popular during the Troubles.

Security provisions in Northern Ireland provided limited opportunities, even though the number and variety of military aircraft had increased. Artist Charles McHugh, himself a Puma 'copter crewman, painted a Wessex (XR529) of 72 Sqn circling Slemish Mountain in Co Antrim. The aircraft ended up a gate guardian at RAF Aldergrove.

a right to be hopeful about the long-term future of the Society.

First came a change in the leadership ranks: In September of 1980, Eddie Franklin stepped down as secretary, to be replaced by Ray Burrows. Eddie shifted his energies and his camera lens to flying matters less restricted—specifically to bird watching. He had been a member for years of the Royal Society for the Protection of Birds, though aviation had been his first love. As son Trevor recalled: "When he was in the RSPB and watching birds, he'd still be listening and watching for aeroplanes."

What the Society lost in Eddie Franklin, it gained in a keen new recruit, one of many swelling the ranks. The

January, 1980 issue of *Airmail* included the first instalment of a series of articles covering the history of Ulster airfields. Researched and written jointly by a number of members, the credits included a now familiar name: Ernie Cromie.
"There was new blood coming in," said Ray Burrows. "They were younger guys who were enthusiasts rather than just spotters. They joined the Society and added something that was missing."

They added numbers to the meetings, for one thing. Helped along by the Society's move to the Ulster Flying Club at Newtownards, attendance rose by 100 percent. But concurrent with an increasing membership was the larger size of the magazine—12 pages being common at this point—and the associated

A trek into the Mournes discovered this wheel from the 1945 fatal crash of a Mosquito. Colin Greer gets his camera ready.

increase in printing costs. Membership fees were jacked up to £5 for adults, but there was a spinoff benefit. Any excess cash from subscriptions would be used to fund guest speakers for the monthly meetings. For the first time in 12 years the members could draw on the aviation experience and knowledge from outside their tight-knit community. So, all things considered, the 1980s had begun well for the Ulster Aviation Society.

Recognizing the increase in activity (and consequent increase in committee workload), Hugh McGrattan humbly offered to relinquish his position as honorary chairman in favour of an 'active' chairman. Ernie Cromie was duly elected and the room passed a vote of thanks for

The Society marquee was a regular fixture at Newtownards' Ulster Air Show throughout the 1980s. Visitors to the show could see the growing collection of retrieved wreckage, learn about the history of local airfields or maybe pick up a surplus copy of **Airmail** *to see what these guys were all about.*

Hugh's tireless work on the management committee in various capacities since the day the Society was founded in 1968.

Ernie inherited an organisation larger and more energised than ever. Visits were high on the agenda, with one to Dublin airport and another with Air Atlantique's DC-3 heading to the Greenham Common display. In May, 1984, the engineering marvel that was Concorde visited Aldergrove, observed of course by the keen spotters of the UAS. The members were no strangers to Aldergrove. Two years earlier, with less stringent but effective security in place, the Society began to organise trips to Aldergrove air traffic control. The organisation also began to attend Newtownards' Ulster Air Show as participants rather than mere spectators. While the crowd were wowed by autogiros and jets above, UAS members met the public down at ground level.

Among the exhibits, visitors might notice a piece or two of recovered aircraft wreckage. Nobody at the time would regard the bits as symbols which would secure the Society's future.

Those who had flirted with the subject of aircraft wrecks in the Society's early years, together with some recent eager recruits, organised themselves in September, 1981 into an aviation archaeology sub-group. This marked a departure in emphasis from aircraft spotting, but as Con

Law later put it: "In those days getting access to the airports, because of the Troubles and the security levels, was not easy… It was a smart move by the Society to get involved in other aspects of aviation like wreckology."

The self-titled 'wreckologists' made their first expedition to Donegal's Bluestack Mountains in search of the wreckage of downed Sunderland flying boat DW110 which had come to grief on the remote slopes on January 31st, 1944. Organizing Secretary David Hill planned the trip, accompanied by Ernie Cromie, Ian Henderson and Robert Wilson. Poor conditions made locating the site difficult but by a fortunate stroke of luck, the team stumbled across the remains. It was the first of several discoveries.

Tuesday evenings at Castlereagh College became the meeting night for the wreckologists, held separately from the greater Society meetings. Here they learned about archaeology, discussed aircraft history and plotted their next expedition, often on the strength of research presented by Ernie.

His skills developed from his study of cartography at university which inspired him to search for the story behind downed aeroplanes. That led him in turn to aviation research sources like the RAF's air historical branch in London.

Ernie Cromie, circa 1980 when he joined the UAS. He served 30 years as chairman and was instrumental in forging the Society from an informal club of enthusiasts into a well-run, formal organisation of over 500 members .

Ernie inherited an organisation larger and more energised than ever.

"And of course, the more one discovers, the more questions are raised," he explained years later. "And the interest just grew and grew."

His knack for investigation proved invaluable and with any luck he would return to his companions with news of a potential crash site, as Ray Burrows recalled: "A lot of us were interested in aircraft crashes, but Ernie was actually

Lynx helicopter XZ666 (nicknamed "Damien", in reference to its treble–six registration) awaits a call to some salvage action with UAS volunteers. They had just uncovered the wreckage of a Rolls Royce Merlin engine at Crockaneel Mountain, Glendun.

Wreckology had become a serious pursuit in the Society in the 1980s. This was an early effort by UAS members to retrieve parts of a Fairey Battle which had crashed on 22 October, 1940 in the Antrim Glens near Cushendall , with two fatalities.

Engine parts are clearly visible in the wreckage of Sunderland DW110 of 228 Sqn, RAF Coastal Command, in the Bluestack Mountains of Donegal, where it crashed in 1944, killing 7 of 13 aboard. The terrain gives an idea of the UAS team's efforts to reach this spot, 2,000-plus feet above sea level.

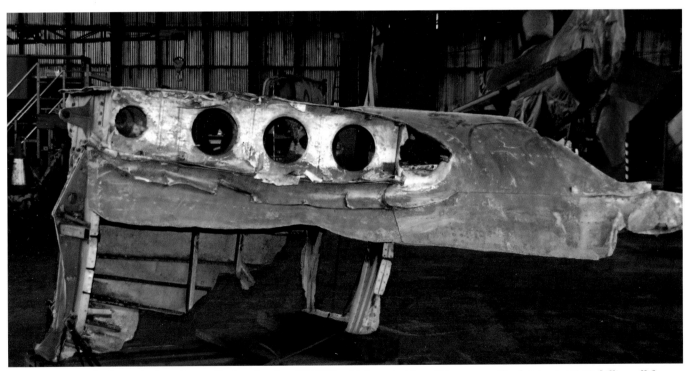

A recovered chunk of rear fuselage from a Hudson bomber (T9328) which crashed into Slievenanee, Co Antrim in 1940, killing all five aboard. The relic now reposes in the hangars of the Society. Its camouflage, while a bit faded, is still in good shape considering its age of some 75 years.

UAS member John Hewitt takes a break at a wreck site on the south slope of Glenariff, in the Antrim Glens. It's the early 1980s, and by this stage the Society's heritage group was gaining considerable experience in accessing such locations in very unfavourable terrain.

going to the Public Records Office, and getting out the files, and coming back with reams and reams of information."

John Hewitt and Cecil Hamilton engaged in their share of research as well, interviewing local farmers and witnesses who could help narrow down the search area. The histories of individual wrecks could prove potentially dangerous, considering that unexploded ordnance might be at the crash site. Fortunately, the search teams never encountered that problem. There could be sensitive matters as well, with the possible discovery of human remains. Again, that didn't occur. The potential was there, however, and that raised the point of an unfortunate incident involving an anonymous member.

He had done a detailed research job and wrote a story for the *Ulster Airmail*, with a graphic description of the bodies. People complained and a private letter arrived at the Society, expressing outrage that such forensic details were disclosed. The member was confronted by UAS leaders, but he defended his action. The matter of sensitivity was secondary, he felt, to the importance of providing the facts, no matter how gruesome. "He was the sort of guy who would not be told," said Ernie Cromie, Society chairman at the time. "If he decided that there was only one way to conclude whatever it was he was investigating, he would just not see reason."

The member quit the Society before he could be expelled.

The wreckology work continued, gathering even greater interest among members. Clambering over rocks and heather-packed slopes in all kinds of weather, the searchers were sometimes able to bring back bits of wreckage from particular sites. Some pieces, such as engines, were simply too large to tote back, but more good luck came their way. Two helicopter pilots from the army's 655 Squadron, Sgt. Steve Shailer and Sgt. Mal Wood, were themselves keen wreck hunters. They shared the enthusiasm with their Ulster Aviation Society counterparts, but they also had something the Society members did not: a Lynx helicopter and a very understanding commanding officer.

In due course visits were paid to various locations to search and retrieve if remnants were found. Soon the dull thud of helicopter rotors echoed through the hills as Lynx XZ666, affectionately nicknamed 'Damien,' would appear in the sky, lifting tackle attached.

Occasionally, the chairman himself joined Sgts. Shailer and Wood to better guide them to their target. On one mission, they helped Ernie aboard and flew up to Chimney Rock in the Mournes to look for remnants of a B-26 bomber. It was quite the trip, he said: "The sides of Chimney Rock Mountain are littered with boulders and the tips of the rotor blades seemed to me to be only inches away from them at times. I was scared witless!"

The wreckologists were gaining momentum and were soon faced with the problem of what to do with their recoveries. Some were kept in members' garages, others dropped off at the Ulster Folk and Transport Museum. There were whisperings of a dedicated aviation museum for Northern Ireland. Then David Telford, UAS member and Transport Museum employee, made an innocent observation: There was a virtually intact Grumman F4F Wildcat which had ditched in Portmore Lough on Christmas eve of 1944.

Membership No. 110

UAS-✈

Hon. Secretary _GWFranklin_

Members Signature _J. Woods_

Jack Woods, who shot thousands of photos for his Shorts employer— and the UAS—was an early Society member. Cards like this gave a semblance of permanence, especially during the 1970s when the Society's continued existence sometimes looked doubtful.

A Symbol for the Society

Ulster Aviation Society members were greeted with a new look *Ulster Airmail* when they opened the latest issue in April, 1981. Inside they would find all the usual features, but leading it was a bold new header, adorned with a striking logo: a winged red hand.

The logo had been designed by member, and soon to be chairman, Ernie Cromie. Though new to the Society, the logo had its roots in the past: "I had a strong interest in military aviation in the Second World War" said Ernie, who was struck by the emblem of America's legendary 'mighty' Eighth Air Force: "their logo was a winged figure eight."

Ernie removed the eight, replacing it with something that would immediately scream 'Ulster' to all who viewed it: "I think a lot of people associated the red hand of Ulster with this part of the world." And thus, the Ulster Aviation Society logo was born, and has remained virtually unchanged to the present day.

2.1 Ulster Airmail

The Society has thrived in its 50 years thanks to the enthusiasm of its members, as well as its fluid reaction to the needs and wants of those members. That flexibility was never made any easier by the prevailing social and political circumstances in its home of Northern Ireland. Throughout all the changes the Society has proudly boasted one constant.

Delays with the Irish Society's Aeronews *often left members with no choice but to write to each other directly on their hunt for the latest gen. Note Eddie Franklin's signature on the bottom right.*

The organisation's monthly journal, the *Ulster Airmail*, was issued for the first time in February, 1969, just two months after the group's foundation. Since then the magazine has been in more or less constant production, issued monthly to every Society member. In fact, Society membership fees are set such that they cover *Airmail* production and no more. All other branches and activities within the Society raise their funds independently. That's how central the *Ulster Airmail* is to the Ulster Aviation Society.

But why have a magazine at all? In 2018, with a membership in excess of 500, it's easy to appreciate its use as a communication tool. But in 1968, with a total membership barely scraping the teens, it seemed a leap to produce a magazine. The fact was that in the late 1960s, communication was slow. The Irish Society of Aviation Enthusiasts had its own periodical, the *Aeronews*, but as Eddie Franklin reminisced years ago, "We slowly became disillusioned with the IAS, as the monthly magazines sometimes ran months late and it was suggested we could do better on our own." Perhaps unbeknownst to the founders, the prosperity of the Society can be directly attributed to the decision to introduce

The hand cranked Gestetner duplicator is primitive by modern standards, but this 1960s era machine enabled the Society to reach members far and wide.

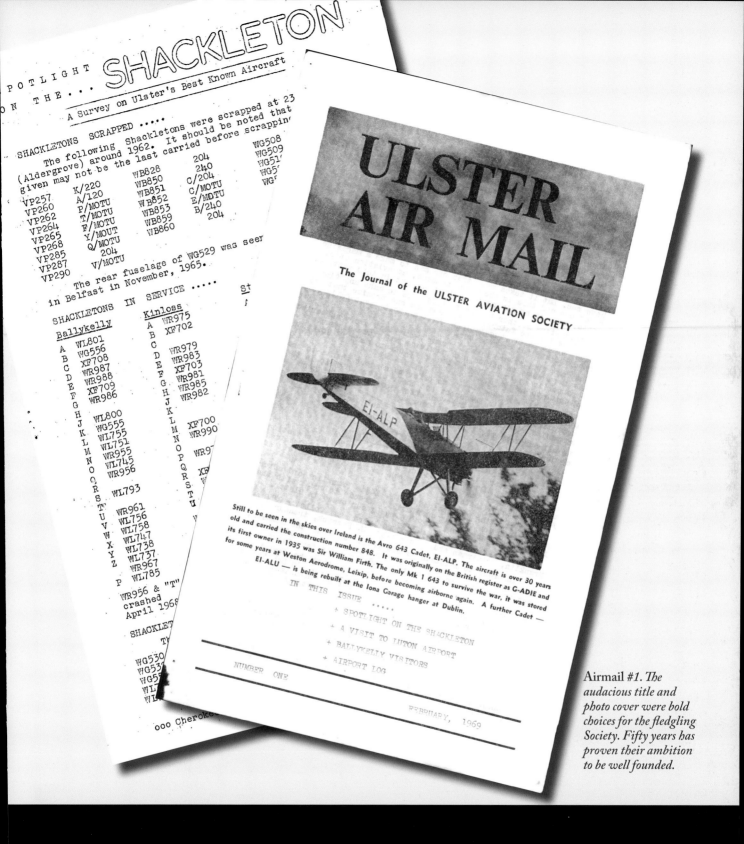

SPOTLIGHT ON THE... SHACKLETON

A Survey on Ulster's Best Known Aircraft

SHACKLETONS SCRAPPED .....

The following Shackletons were scrapped at 23 (Aldergrove) around 1962. It should be noted that given may not be the last carried before scrapping

VP257	K/220	WB828	204	WG508
VP260	A/120	WB850	240	WG509
VP262	P/MOTU	WB851	C/204	WG51
VP264	T/MOTU	WB852	C/MOTU	WG5
VP265	F/MOTU	WB853	E/MOTU	WG
VP268	Y/MOUT	WB859	B/240	
VP285	Q/MOTU	WB860	204	
VP287	204			
VP290	V/MOTU			

The rear fuselage of WG529 was seer in Belfast in November, 1965.

SHACKLETONS IN SERVICE .....

Ballykelly		Kinloss		
A	WL801	A	WR975	
B	WG556	B	XF702	
C	XF708	C		
D	WR987	D	WR979	
E	WR988	E	WR983	
F	XF709	F	XF703	
G	WR986	G	WR981	
H		H	WR985	
J		J	WR982	
K	WL800	K		
L	WG555	L		
M	WL755	M	XF700	
N	WL751	N	WR990	
O	WR955	O		
Q	WL745	P		
R	WR956	Q		
S	WL793	R	X	
T		S		
U	WR961	T		
V	WL756	U		
W	WL758			
X	WL747			
Y	WL738			
Z	WL737			
	WR967			
P	WL785			

WR956 & "T" crashed April 1968

SHACKLET

WG530
WG53
WG5
WL
WL

ooo Cherok

ULSTER AIR MAIL

The Journal of the ULSTER AVIATION SOCIETY

Still to be seen in the skies over Ireland is the Avro 643 Cadet, EI-ALP. The aircraft is over 30 years old and carried the construction number 848. It was originally on the British register as G-ADIE and its first owner in 1935 was Sir William Firth. The only Mk 1 643 to survive the war, it was stored for some years at Weston Aerodrome, Leixip, before becoming airborne again. A further Cadet — EI-ALU — is being rebuilt at the Iona Garage hanger at Dublin.

IN THIS ISSUE * * * * *

+ SPOTLIGHT ON THE SHACKLETON

+ A VISIT TO LUTON AIRPORT

+ BALLYKELLY VISITORS

+ AIRPORT LOG

NUMBER ONE

FEBRUARY, 1969

Airmail #1. The audacious title and photo cover were bold choices for the fledgling Society. Fifty years has proven their ambition to be well founded.

a magazine so soon after formation, because that magazine gave the Ulster Aviation Society something unique: reach.

The majority of Society spotters gathered at Aldergrove, and the subsequent meetings were held initially in a variety of locales in Antrim—not very accommodating to the remainder of Ulster for whom the Society was named. But with the issuing of *Ulster Airmail*, the Society could suddenly reach as far afield as it wished, allowing those who couldn't frequent Antrim to feel part of the group nonetheless.

Nowhere was this more evident than on Ulster's north coast, home of founding member and later chairman Hugh McGrattan. Hugh was nowhere near Aldergrove and not a regular meeting attendee, but in *Airmail* he saw another avenue by which he could get involved, so he volunteered with boyish enthusiasm to take on the post of *Airmail's* first editor.

"There I was, the smile of one who has just discovered a Whitley buried in his back garden on my face, and in my hand a title page," he wrote in an early edition.

The magazine soon faced its first hurdle when in 1970 Ulster Air Transport, which had graciously printed *Airmail* on behalf of the ragtag bunch, folded. The day was saved thanks to young member Con Law. He lobbied his employer, Woodgate Aviation, to allow the Society to use its Gestetner duplicator after hours.
The Gestetner was a cumbersome machine, with each page being individually stencilled then duplicated with a hand-wound crank. The smell of wet ink filled the air. But for the growing organisation it was a lifeline, so much appreciated that when Woodgate Aviation upgraded, the Society bought the old Gestetner for its own use.

When Hugh McGrattan stepped down as editor in January of 1972, Roger Andrews stepped up, blissfully unaware that his tenure would see the magazine through the most difficult years of its history. The Society languished during the Troubles, cut off from the beloved airfields and airports by strict security regimes. "Aldergrove is taboo, Ballykelly is barren," lamented Eddie Franklin. The magazine suffered in sympathy. In the mid-1970s it was still largely a journal of spotters' logs. With spotting heavily restricted, the logs stopped coming. "There was always a plea for people to provide information," Con Law recalled. "There was in the mid to late '70s a problem where you thought, 'Are they going

The Society used the Ulster Air Show to meet and greet the public. Note the copies of Airmail *behind Ray Burrows – what better way to introduce a potential recruit?*

By the early 2000s, Airmail had changed almost beyond recognition with the move to commercial printing in 1994, and the introduction of colour in 2001.

ULSTER AIR MAIL

The Journal of the

CHAIRMAN
H.G.McGRATTAN
8 Rodney Street
Portrush
Co.Antrim.

ASS.EDITOR
C.LAW
23 Broughshane Rd
Ballymena
Co.Antrim

EDITORIAL......The first for som
Season is with us again and it
shows you happen to see. You
at Newtownards on the first
information you can now cont
it) at Antrim 3238, he will
interest for the magazine.

LOCAL NEWS.....SD.3-30 c/
has been seen as G-BETN
G-BDSU left Sydenham on
was in America in May
to Swinderby for this
1954) will become JA
French Navy Alouett
on May 28th.
Some ownership ch
with Davidson and
owns the Nipper
The remains of
to our Good Ru
-craft at Alder
A visitor to the Fly
resident/visiting aircra
G-AWUU, G-BAAT, G-AVUX, G-BAB
269 EI-APN, Robin {**} G-BAEC.
Autocrat G-AJEE. Travelers, G-BBBE,
G-AXLI. Twin Comanche G-AVHZ,

Spitfire 'ENNISKILLEN'

VOLUME NINE, NUMBER SIX

ULSTER AIRMAIL

The Journal of the Ulster Aviation Society

Vol. 26 No. 4

ULSTER AIRMAIL

The Journal of the Ulster Aviation Society

VOLUME 33

JUNE 2001

FLT LT H Smyth from Northern Ireland demonstrates Harrier GR7 ZG862 to the crowds at the City of Derry Air Show on June 3rd (Photo P Harvey)

SYDENHAM 1941-45
AIRSHOW REPORTS
AIR ACCIDENT REPORTS
CIVIL AND MILITARY NEWS
PLUS ALL THE REGULAR LISTINGS
NOW WITH LOTS MORE PHOTOGRAPHS!!

to be able to continue, have they got the funds to do it?'"

Cries of desperation from the editorial team resounded through the membership and by the early 1980s the *Airmail* had returned from the brink. Among the enthusiastic new contributors was future chairman Ernie Cromie, one of the originators of a project to record the history of Ulster's airfields. Historic articles weren't new to the magazine, in fact *Airmail's* first year saw stories on the Lockheed Corporation, the 81st Tactical Fighter Wing and home-grown hero Harry Ferguson sharing the pages with up-to-the-minute aircraft movements.

But "Ulster's Airfields: A Complete History" was to be the first long-running, serialised feature ever published in *Airmail*.

Recurring features weren't limited to heritage interests. As the art of aircraft spotting evolved, its practitioners made the most of *Ulster Airmail* as a platform for sharing the latest developments of their favourite hobby. In the late 1970s aircraft spotting via amateur radio was becoming an increasingly popular spin on the pursuit and *Airmail* contributors were quick to update their readers of the intricacies of the new-fangled technology. May 1983 saw the debut of the enigmatic Ray D. O'Shack. His regular columns were sprinkled with doses of deadpan humour to leaven the serious content: "My present task is to ensure the crowd fencing will be electrified at the forthcoming Ulster Air Show to keep the unruly elements who featured on the Dublin trip in check." Portable radios were making their mark among spotters, and he provided helpful advice: "For best results, a scanning type radio is recommended because it is, in effect, several radios in one." O'Shack's column gave way to Ron Bishop's "Airband Information" in 1986, which remained a feature for many years, growing from a few lines squeezed into a jam-packed magazine to several pages of logged aircraft, rivalling the airport spotters' notes for space.

Stephen Boyd picked up the editorial reins in 1986. A growing membership had put pressure on the *Airmail* production team, with each issue now photocopied by members such as Marc Steenson or Ian Logan and delivered to the Society's management committee, which concluded its monthly meetings by collating and stapling each of the A4 issues. It was mailed to a membership of almost 360 and by 1994 committee members were spending more time preparing the *Airmail* than discussing Society matters.

Editor Graham Mehaffy's work is almost entirely digital. But as well as editing he contributes a monthly editorial and frequent meeting reports to the publication.

2.1 | Ulster Airmail

"We'd have many an end to a committee meeting where we went around the table, picking one page up after another, and the final person stapled the magazine," recalled Stephen Boyd of the frustration of issuing *Ulster Airmail* month after month. The time had come for a change.

Stephen's legacy as editor was to establish the commercial printing of the *Airmail* by Belfast printers Northern Whig, relieving the committee of the tedium. The new format *Ulster Airmail* dropped through members' letter boxes in April, 1994. Now A5 in size and stapled along its spine, it looked like a true magazine for the first time in its history, departing from the corner-stapled newsletter version of years past.

When Gary Adams was introduced as editor in January 2001, he wasted no time in making his mark. A focus on Society news and the copious use of photographs were hallmarks of Gary's publications, but his editorship may be best remembered for the introduction of full colour front and back covers, a quantum leap from the hand-cranked, ink-stained copies of 1969.

When Gary stepped down in 2003 Michael Bradshaw graciously volunteered as acting editor but restrictions on his free time were painfully evident, with magazines being combined and deadlines missed. Graham Mehaffy recognised a publication in peril: "(Michael) said, 'I can't do this anymore', so Ernie was really worried." With the Society in the midst of the Langford Lodge tenancy crisis, it was a headache the chairman could do without. Graham took on the post in April 2005 and, as of the 50th anniversary, remains the editor of *Ulster Airmail*, his background producing posters and leaflets for the Ordinance Survey standing him in good stead.

Today's magazine is a world apart from the eight-page newsletter that Hugh McGrattan held proudly in his hands in the early months of 1969. Now full colour throughout, the content ranges from historic articles and meeting reports to Society news and members' memoirs.

Graham's magazines are notable for their variety and he doesn't limit articles solely to local interest pieces.

"So as far as I'm concerned, if somebody goes to a museum in Paris, there's a good chance that somebody in the Society would say, 'That looks interesting; I'm going to Paris next year, I'll go and visit it,'" he explained. "That's the way I look at it; I don't want to be too parochial. And I can't be anyway. Thirty-two pages to fill every month?!"

As for the challenge of editing the magazine: "Yeah. I enjoy it ... Sort of stretches your brain a bit."

Struggling spotters, exhausted editors and printing pandemonium have each challenged the *Ulster Airmail* in its 49-year history, "... but they got through," smiles Con Law, "and I always look forward to the magazine dropping on the doormat every month, to this day."

And the title? Thank inaugural editor Hugh McGrattan: "Oh yes, I also gave the magazine its name."

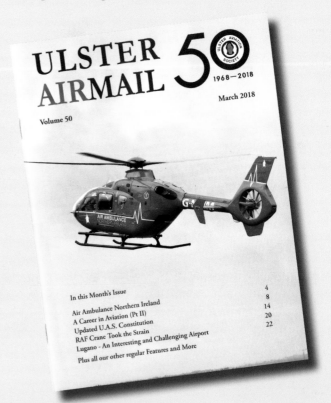

Society members look can look forward to a high quality, full colour, professional publication delivered direct to their door (or digitally via email) every month of the year.

Today's Ulster Airmail *is printed by Peninsula Printing in Newtownards. Every month over 500 copies are produced before being distributed to members by Paul Trimble Printing.*

CHAPTER THREE

High speed: The UAS is familiar with that performance quality. It's basic to the design of so many aircraft, some of them in the organisation's own modest collection.

So members nodded in appreciation on October 4th, 1983, when Scottish entrepreneur Richard Noble, pilot and inventor, fired his Thrust 2 jet-powered car to a 634 mph speed record on a sun-baked Nevada desert.

Speed, a modern pleasure. Noble's run was a counterpoint to the reserved pleasure of patience which some of those same members embraced the following day at another

Most of the carcase was below the surface, snug in a bed of soft silt.

flat surface: a tiny, marsh-bordered lough, 5,000 miles from Nevada's racetrack exuberance.

A Lynx helicopter from 655 Squadron, Army Air Corps, with rather more discreet jet power, hovered above a wee cluster of UAS volunteers awkwardly afloat in a puny, home-made punt on Portmore Lough. Like Noble, they were about to make history on a modest scale—more local but

no less proud.

Beside them, its broken tail barely poking above the water, was a Second World War fighter aeroplane. What could be seen of it was blistered, battered and spotted with gull dung. Most of the carcase was below the surface, snug in a bed of soft silt.

On December 24th, 1944, Grumman Wildcat JV482 of the Royal Navy's Fleet Air

JV482 in Portmore Lough, the remains of its tail in the foreground. It was a wreck, but some Society members could see its potential.

Arm, had suffered an engine fire minutes after take-off from RAF Long Kesh airfield. The young pilot quickly ditched the fighter, which sank five feet, then stopped. It had reached bottom. Its occupant climbed along the spine, waited for a civvy rowboat to rescue him and returned to Long Kesh in time for his Christmas Eve dinner.

The Wildcat sat where it was for 39 years, its visible bits a target for souvenir shredders and rude birds—that is, until the UAS arrived on the scene in October of 1983. A mere six weeks had passed since a suggestion had been posed to retrieve the aircraft. The doubts, ideas and planning made a hefty meal for the amateurs but, once agreed, they slaved in the days that followed: They would retrieve this half-sunken hulk. Jim Walsh, superintending engineer from George Heyn's heavy lifting business, would oversee the operation with assistance from Society volunteers and the Belfast Sub-Aqua group.

It was decided that only a helicopter could safely manage a secure and effective lift. Up stepped Sergeants Shailer and Wood once more with their Army Air Corps Lynx.

"They were mad keen to get that Wildcat out of Portmore Lough," said Ernie Cromie. Workers hauled the propeller, then the corroded engine free from the rest of the wreck and lowered them to an onshore bed of grass.

Engineer Jim Walsh (right) oversees removal of JV 482's propeller using his specially-constructed frame. Recovery of the Wildcat could not have been achieved without his leadership.

A detailed survey of the wings and fuselage left no doubt they would require a more careful approach, and months to put into effect. But, once accomplished, this would be more than just a salvage victory. It would represent a fundamental change in the aims and work of the Ulster Aviation Society. "It was the recovery of the Wildcat which resulted in us finally making up our minds irrevocably and firmly that it would be the most important basis of an aviation museum," said Ernie Cromie, looking back many years afterward.

The UAS annual general meeting in 1984 provided the most interesting election in the Society's history, with Ernie running for chairman against two other stalwart members:

Position	
Hon. Chairman	
Proposed	Votes
R. Burrows	6
E. Cromie	13**
E. Cadden	2

Three candidates for the 1984 chairman's position was a clear sign of an active, healthy Society.

Ray Burrows and Eddie Cadden. Ernie secured the majority—a testament, perhaps, to the Society's progress since his election as chairman two years earlier. With his new mandate, he drove forward a personal challenge to provide the organisation with a formal constitution. Simply put, it might make money-raising simpler. The Society's growth in the early 1980s had posed a problem: the new projects undertaken by the volunteers had to be funded somehow. The management committee—itself not a legally formal entity—had rarely focused on income or other financial formalities. After all, the printing of the *Ulster Airmail* had been the only significant cost and membership fees were structured to cover only that expense.

The days before health and safety limits: Society members gather beneath the 655 Sqn Lynx as the Wildcat's engine is delivered.

DRAFT CONSTITUTION

NAME

The name of the organization shall be the Ulster Aviation Society, hereinafter referred to as 'the Society'.

OBJECTS

1. To assist members of the Society in furthering their interests in particular aspects of aviation and to provide them with a forum for discussion of matters of mutual interest.

2. To liaise with other similar organizations and aviation interests generally and to co-operate with these bodies on matters of mutual concern.

3. To act as a centre for giving or obtaining advice and information in order to foster awareness of all aspects of aviation, past and present.

4. To work for the establishment of an aviation museum for Northern Ireland.

5. To arouse public opinion in order to further the promotion of these objects.

For the first time, the purpose of the Society was laid out in black and white. Not all agreed with it.

A friend of Ernie planted the seed with a basic challenge: "Do you people have a constitution? Don't you think it would be beneficial if you were looking for financial assistance? A constitution would stand you in better stead.'"

He had a point, one that some Society activists could not appreciate. Why formalise a group which had survived through tough times? And anyway, were they not a spotters' organisation? All this effort on wreck hunting and aircraft recovery was, in the eyes of some, a distraction at best and a catastrophic change of direction at worst, evidence of museum dreams and the like, too grand for a bunch of lads who simply liked aeroplanes.

The diehard spotters felt threatened. They strongly opposed a document which would render wreckology or museums a core part of the Society's purpose—and perhaps by default minimise the importance of spotting.

They made their point by rejecting the constitution proposal. The constitution's supporters, with Cromie and Burrows in the vanguard, went to work. After nine months of gentle persuasion and guarantees to the spotters, the document went to the annual general meeting. This time, it passed. Formalising of the Ulster Aviation

Success! After 40 years in Portmore Lough, JV482 takes to the air again with a bit of help.

The UAS was keen to show off the new prize. Visitors to the 1984 Ulster Airshow are fascinated with the Wildcat, a mere three months after its recovery.

Society management and its operations began in earnest. Debate among the membership was one thing, but a stalemate within the committee would not be easily healed.

Graham Mehaffy, a committee member at the time (and later longtime editor of the *Airmail*) recalled the atmosphere: "Some members of the committee felt the focus was moving from the Society itself and the running of the Society to the museum project....It nearly came to blows some nights."

That was no exaggeration, according to some Society members who witnessed the discord. The issue bubbled

This painting by David Moore depicts JV482 in its role as a Royal Navy fighter before the Portmore crash ended its career. Below is its seaborne base, HMS Searcher.

beneath the surface for years, an irritant which refused to vanish. Undeterred, the Society at large soldiered on. Typical was the determination of the wreckology group.

Their recovery of the Wildcat fuselage was never going to be easy, but after weeks working to lighten and float the hulk, the salvage crew hauled it from the lough sludge on April 30th, 1984. Slung beneath its Lynx "rescue wagon," the Wildcat took to the air for the first time in 40 years and was lowered onto the shore. The final effort was thanks again to the seemingly boundless enthusiasm of Sergeants Shailer and Wood, and the expert guidance of Jim Walsh. It still may be the single most momentous undertaking of the Ulster Aviation Society.

Less than two weeks after its recovery, JV842 was paraded proudly at the Belfast Lord Mayor's spring show and again on June 2nd at the Ulster Air Show. The aircraft, battered and bruised though it was, made quite an impression on onlookers, and certainly stood out as unique among the static exhibits.

The Society seemed seized by a new dynamism at this stage, engaged in a wide variety of activities. On June 21st, a group headed south through Dublin to Shannon Airport for a weekend stay, with spotting, a control tower tour and a five-star lunch—the latter as guests of logistics business SRS Aviation.

The gun camera from JV482, used to complement post-combat reports.

Further successful trips followed including the Fairyhouse Air Spectacular north of Dublin in August and a tour of Dublin Airport for 32 members in October. The growing respect for the Society was evident in Ernie Cromie's invitation to address the distinguished Royal Aeronautical Society on the subject of Northern Ireland's military airfields that same month.

Meanwhile, the wreckology activists—now known as the heritage group—had successfully removed the wings from the Wildcat. They were stored in a lean-to shed at Castlereagh College along with the fuselage. Inside a school annex, volunteers dismantled the engine. It was a start, but the full restoration would take many years to complete.

Little had been planned beyond the recovery success. However, that achievement alone had given impetus to the dream of establishing an aviation museum. That grand goal had even become entrenched in the new constitution.

Meanwhile, the Society's heritage group continued spreading the word about the Wildcat project. Ray Burrows was part of that action and later in 1984 his efforts brought a delightful result from one of his letters—a moment he's never forgotten.

"I actually got a phone call one day from a guy who said, 'I see you're looking for the pilot of the Wildcat that crashed…I know who it is…I was flying with him that day'!" The pilot of JV842 was Canadian resident Peter Lock and, after some long-distance digging, contact was made with the retired aviator.

Ray broke the news of the raising of the wreck from Portmore Lough: "There was stunned silence; he was absolutely gobsmacked." Once Peter could muster the words, he told Ray that, as it happened, he returned to the UK every winter. This one would be unique. On January 7th, 1985 an eager Peter with wife Marjorie touched down at Belfast International Airport. A nervous but jubilant bunch of Ulster Aviation Society members welcomed their special visitors, and national news teams joined in, keen for a feelgood story in an otherwise troubled Northern Ireland.

Peter and Marjorie proceeded on a whistle-stop tour of the province. It took them first to the homes of members of the Kane family. Seamus and William Kane had rescued Peter from his ditched fighter back in 1944. Then it was on to Castlereagh where Peter was greeted with the sight of the recovered Wildcat—his Wildcat. A somewhat emotional Peter graciously presented the Society with his log book, including the entry for Christmas Eve, 1944. When he left Northern Ireland, it was with a promise to return, even if just to check on his Wildcat's state of restoration.

The work was slow, but the Society by 1985 was in motion on other fronts. Trips continued with over 30 members once again enjoying a day's spotting at Dublin on the occasion of an Ireland v France rugby match. Legend has it that the craic was such that one Society member spent most of the following day in a wheelchair—and not on medical grounds! Perhaps the hospitality at Dublin Airport was too good to resist. Although the year seemed quiet in comparison to the landmark that was 1984, John Hewitt had an ace up his sleeve. The Society hosted speakers, organised trips, recovered historic artefacts and published a monthly magazine—all quite impressive for a community of volunteers. But John Hewitt's idea was something new. On December 15th (by coincidence, the Society's 17th birthday), a collection of members and friends gathered at their 'home turf' of Aldergrove and one by one ascended the steps of Hawker Siddeley Trident 3 G-AWZU, soon to be retired. John's successful campaign saw the group purchase enough seats to hire the aircraft outright for a one-hour flight. Even better, tickets cost only £35 per person. And to top that again, the Ulster Aviation Society was allowed to select the route! The aircraft hurtled down the runway and rose with ease into the sky before heading south on the first leg of a breakneck

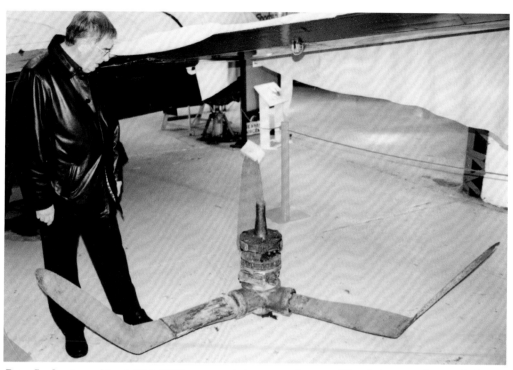

Peter Lock returned to visit his Wildcat several times over the years. Here he inspects the wrecked propeller, bent from the force of the crash.

The measure of the man: Peter Lock summed up the engine fire and subsequent ditching with just 17 words in his 1944 log.

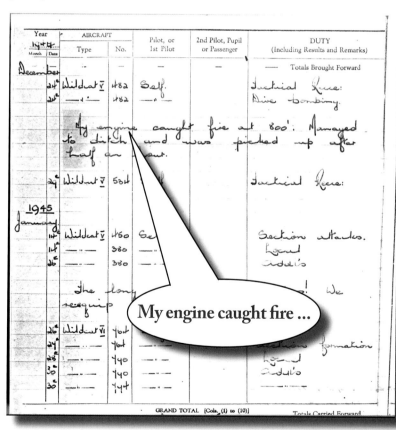

tour of the province. Banking left and right, first high altitude then low, the Trident hauled its eager cargo in all directions to sights of special aviation interest as they munched and sipped their way, sometimes with more gusto than good sense. What a way to end a year, as Ray Burrows remembers: "Did we enjoy the trip? I don't think anyone on board will ever forget it!"

The Trident flight had whetted appetites which led to other aerial adventures. Stephen Boyd, trip organiser supreme, courted Air Atlantique for four months until, on September 27th, some 180 members took their turns on one of five short, sellout flights in a DC-3, icon of air transport.

Back on solid ground, members by now had begun taking greater interest in the aviation scene around Newtownards. The airfield and the Ulster Flying Club had been familiar territory to the Society for meetings and air shows since 1980, with a highlight of 1985 being the arrival of a Wessex helicopter of 72 Squadron. Aboard was Sub Lieutenant Paul Critchley, landing in style to provide a talk on air/sea rescue for a monthly meeting. At another gathering, Shorts test pilot Alan Deacon drew a record crowd of 55. By now, Society members were welcoming a wide variety of fascinating speakers in the Flying Club meeting room. But just down the road in Newtownards, a more exciting development had begun to take shape.

Peter Lock was airborne less than a week after the ditching ordeal on Christmas Eve, such were the pressures on wartime airmen.

Ards Borough Council was showing increased interest in Society activities, abetted by regular media attention which had increased as a result of the Wildcat salvage. In December of 1985, the council sent the UAS a Christmas present in the form of a letter which arrived at the door of Chairman Ernie Cromie. The council had taken note of the Society's long-term goal of a museum. It wanted to be part of the action. The idea simmered for a while, nursed along with informal chats and sketchy plans for a formalised collection—perhaps on the grounds of Ards Airport. Both parties nursed visions of a home for aircraft, displays and workshops.

By 1986, with the Ards council approval, the Society's heritage group had shifted the Wildcat to an abandoned abattoir near Ards Airport. Ernie's initial reaction was positive.

"Absolutely delighted," he said at the time. There was room for work on the Wildcat, and a bit extra besides. It was basic preparation for what the heritage group figured could be an eventual move to a museum, to be sited on the grounds of the airport. The slaughterhouse consisted of four walls, a mild odour and very little else. Ray Burrows, years later, admitted he was disgusted when he first laid eyes on the building: "What we didn't realise was that half the roof was missing… and there was about 12 inches of dried blood on the floor."

Members had to clean the place up and get power to the Wildcat to run tools, further hampering an already laborious and time-consuming restoration, so the team took solace in the proposed museum. After all, the abattoir was to be nothing more than a stopgap until the museum project took off. There would be many complex

JV482 arrives at Castlereagh College, the first of at least four locations to become home.

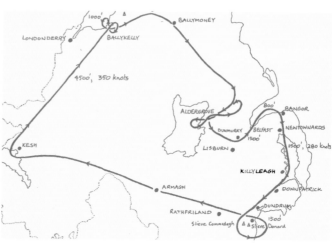

Between driving, flying, spotting, touring and photography during the 1980s, members enjoy a few minutes for refreshments. Left to right are Alan Watson, Stephen Boyd and Bobby Bilsland. Les Thompson looks on from the table behind.

Value for money: Their single flight took the members on a whirlwind tour to every corner of the province.

and challenging steps toward that goal. And yet the museum vision was only a limited part of what the Ulster Aviation Society was doing. There was a magazine demanding monthly attention, finances to arrange, air shows to visit, stories to write, a media beast to feed, regular membership meetings and guest speakers to book. There were other aircraft to acquire, transport and restore. And it's worth noting that, as membership has grown, only a diligent fraction of members have been actual volunteers for many of those activities.

Society members descend the steps of the Trident after their sellout flight. It was the first of many flight experiences organised by the UAS.

ER AVIATION SOCIETY in conjunction with AIR ATLANTIQUE
presents*

ASURE FLIGHTS—£30 PER SEAT

ht of approximately 40 minutes duration aboard the famous Douglas DC-3 Dakota

For further information telephone Belfast 616659 or 790699
or send £30, payable to ULSTER AVIATION SOCIETY to
Boyd, 7 Malone View Road, Belfast BT9 5PH or Jim Redpath, 2 Orby Parade, Belfast 5.

ON SATURDAY 27TH SEPTEMBER 1986

Pleasure flights were becoming a regular Society activity. The DC-3 was a legend, and members jumped at the chance to fly in one. It visited Belfast for five sold-out pleasure flights organised by Stephen Boyd.

In November of 1986, the chairman of Shorts knocked at the door. Sir Philip Foreman, obviously impressed by the Society's aims and energy, had an aeroplane to donate. It was a Shorts SD3-30 prototype, a twin-engined star of the commuter airlines market.

The handover of the aircraft was a resounding endorsement of the museum project, since at the time of the donation the Society had no facility in which the sizeable aircraft could be stowed. It was agreed that G-BSBH would remain in storage on the airfield at Sydenham until such time that the Society could accommodate it.

The significance of a major manufacturer such as Shorts

> ## Sir Philip Foreman, obviously impressed by the Society's aims and energy, had an aeroplane to donate.

handing over a prototype of the successful SD3-30 to a group of volunteers and amateurs cannot be overstated. Respect for the Society had grown, and word of a museum had spread.

Initial discussions with the Ards council were met with some trepidation among the Society members themselves. If the project was to succeed, the collective membership would have to be won over. The hesitation of some members continued to hang on the notion that the Society was losing sight of its roots. What had begun as a casual spotters' club was now a constitutionalised organisation seemingly bent, they figured, on collecting aircraft and building a museum. Beyond the dubious principal of

such a venture, there were the associated expenses. It cost nothing to spend a day at Aldergrove, watching the aircraft come and go. As for trips, members paid for them from their own pockets, and those who didn't go, didn't pay. The heritage projects were a different matter. They were very expensive in comparison. The fear among those who had little interest in the subject was that their own hard-earned money would be funnelled toward Wildcat restoration or another heritage mission. To meet that concern, the management committee endorsed a Wildcat Fund, set up in 1984. Members could donate to it directly, whereas funds raised through membership fees would not be spent on restoration

Newtownards Abattoir: The team weren't too pleased with the conditions here, but the upcoming museum project had hopes running high.

projects. This went some way to appeasing those who were sceptical of heritage schemes, though it did little to alleviate the concerns about the direction in which the Society was headed.

With the membership cautiously on board, the next step was to determine the viability of a museum. Ards Borough Council asked the Society to submit a proposal. It was ready by July of 1987. The initial capital cost discussed was £658,000 for a simple aviation heritage institution. It would house nine aircraft, plus displays, workshops and such. Discussions appeared to give the Society almost complete control over the direction of the project, with the council willing to help financially if it proved workable. There might be European grant funding available as well, to ease the burden. Progress at the political level, however, was slow. Ards had no expertise in the design or construction of a specialised aviation museum, and the Society itself had no experience in running one. So the council asked a private firm, L & R Leisure, to review the plans.

While the consultants' museum research work got underway in the background, the Society continued to grow. A relationship was blossoming with management at Dublin Airport, nurtured by Stephen Boyd and Dublin member Antoin Daltun and frequent visits continued, strongly

Though donated to the Society, circumstances would see G–BSBH end its life on the scrapheap. Its flying compatriot, Shorts Sandringham 'Southern Cross', had a happier fate at the Solent Sky Museum in Southampton.

The restoration of JV482 required tools, materials, accommodation and power. The expense of the lengthy project drew criticism from some corners of the Society.

supported by the membership. The committee, now well aware of the need to fundraise, took every opportunity to bring in those extra few pounds. The casual visitor to the Ulster Aviation Society tent at the 1987 Ulster Air Show could find for sale tea towels, postcards, posters, prints, etchings and model kits. The successes of the preceding years, coupled with increased media attention, paid dividends for the enthusiasts, with membership passing 200 for the first time that July.

The air experience flights proved popular and the growing relationship with Aer Lingus and Dublin Airport was explored to its full advantage.

On May 22nd, members

"Ernie, do you want a Vampire?"

enjoyed an unforgettable flight on Aer Lingus Shorts SD3-60 EI-BPD from Dublin to Shannon. The unforgettable bit? Ray Burrows recalls: "What we were not expecting was a touchdown at Shannon in a 25-30 knot crosswind!" The less than graceful lines of the SD3-60—high wing and slab sides—do little to dampen the effects of a strong wind on left or right flanks.

The monthly meeting attendance continued to grow; typically, the more prestigious the presenter, the bigger the audience. That was certainly the case for the September 1988 meeting at which Wing Commander Ron Shimmons enraptured the crowd of over 100 with his knowledge of the Tornado FG1 and the

Battle of Britain Memorial Flight. It would prove to be a promising visit as well. Much impressed with the work of the Society, Wg Cdr Shimmons, head of operations at RAF Coningsby, made a phone call to Ernie Cromie in early November. "Ernie, do you want a Vampire?" the wing commander barked down the line. Taken aback at the surprise call, and it being mere days since the Halloween holidays, Ernie later confessed to some initial confusion as to the nature of the 'vampire' that was on offer! "It took a few seconds for the penny to drop" he admitted.

Wg Cdr Shimmons, an Ulsterman himself, was offering the Society a complete De Havilland Vampire

Visitors to the UAS stall at the Ulster Airshow could buy books, videos and artwork. The proceeds went some way to funding the heritage projects.

aircraft. Neither the Society nor RAF Coningsby wasted any time and by January, 1989 the twin-seat jet trainer was nestled at Sydenham, albeit at a cost to the Society of around £800.

The committee was suddenly under pressure to seriously consider the future of the heritage collection and how best to manage it. The museum project was maintaining its momentum at the time and hopes were high that soon the collection would have a place to call home. In the meantime the Wildcat was located at Newtownards, with the SD3-30 and Vampire at Sydenham. Plans were underway for the retrieval of a Sea Hawk fighter from Shorts. Money was the issue of the day, and until the museum project got the final go-ahead with the full backing of the council, the Ulster Aviation Society would have to bear the hefty financial burden of the heritage collection alone.

"That caused a few arguments," former committee member Graham Mehaffy recalled. "Why was this money being spent on recovering bits of aircraft when it should be spent on the Society as a whole rather than that particular aspect of it?"

Arguments, said Graham. Occasionally "notably acrimonious" arguments, confirmed former committee member Stephen Boyd, a resolute spotter, then photographer and a long-serving UAS member.

To meet the concerns of some members, the Society launched the Heritage Investment Project—a direct debit style scheme through which members could contribute a monthly sum to the Society to fund the increasing work on a museum collection. In a similar vein to the earlier Wildcat fund, it allowed those members who had an interest in history and restoration to contribute towards it, while separating heritage from the main Society accounts. It would ensure that other activities weren't sidelined.

The museum consultants published their museum report in November, 1989. It looked grim. Under a guise which seemed positive at first blush, L & R grandly expanded the Ulster Aviation Society's idea of a mere aviation heritage museum.

Marketing was high on the consultants' list of priorities. The Society's proposal would cater only to people who already were aviation enthusiasts—and most of them would be men, it warned. It was women who decided on days out for families, it noted. The UAS plan would not create "sufficient pull" to attract a significant portion of the general public, it said. L & R suggested a highly-enhanced version of a museum. It was bound to be expensive. An enterprise—which it dubbed "The Flight Centre"—for enthusiasts, families and casual visitors would provide wider market appeal, it said.

The broad definition, in

"Why was this money being spent on recovering bits of aircraft when it should be spent on the Society as a whole rather than that particular aspect of it?"

retrospect, was grandiose — even somewhat bizarre, in the Society's eyes. Flight, under the original proposal, specified aircraft. But L & R, in order to expand appeal, also saw room for birds, on account of the ornithological significance of Strangford Lough. In fairness, L & R's proposal made no mention of bats or flying fish. It did suggest more interactive exhibits for children—a positive aspect, and one in which mothers had the greatest influence, said the report.

Finally, in order to meet costs, the Society would have to welcome about 120,000 visitors a year to the flight centre. The report itself and other sources didn't say so directly, but that level of attendance seemed doubtful. The report's bottom line was the financial shocker: "The Flight Centre would involve capital costs in the order of £2.5 to 3 million."

The Society's committee was not impressed. It submitted a counter-proposal which incorporated prefabricated buildings as a cost-cutting measure.

"We had worked out, using sectional buildings, that we could build something substantial to house our collection for just over £1.1 million," said Ray Burrows. That, however, at almost a third of the L & R figure, would not buy the type of project envisaged by the consultants. In the months which followed, there was talk

The Vampire arrives at Sydenham. The Society's expanding collection would soon fill the halls of a bespoke museum, or so they thought.

of grant schemes and fund raising, but within the Society there was little heart for the L & R concept.

Ernie Cromie summed it up when he met with Ards Borough Council in October of 1991: "We have spent the last seven years moving toward the project, which today seems further away than ever."

Not only far away, but vanishing over the horizon. The consultants' report had induced a creeping chill in the council. Despite a brief frisson of hope with the production of some optimistic drawings,

> "We have spent the last seven years moving toward the project, which today seems further away than ever."

the politicians' original ardour gradually withered away.

"I knew the council were going to be up against it," Ernie said years later. "Ards Council was one of the smaller councils in Northern Ireland. They simply did not have the wherewithal to generate the amount of money that was going to be required. That was, of course, extremely disappointing."

In retrospect, it was perhaps just as well. There was still a Society at large to support, and that would require money—albeit in more modest amounts. The abattoir

was falling down around the Wildcat, other restoration projects awaited attention and there were members' trips on the books.

In the years while the museum/ flight centre discussions teetered, Society members had enjoyed other diversions. A talk by legendary pilot Roland Beamont in 1989 drew an audience of 80. Such events were becoming more frequent. As well, air shows in the U.K. remained popular among Society members. Nearly 100 had gone to the Prestwick air show that June. That tour alone powered hangar

The Sea Hawk formed the centrepiece of Shorts' training school for over 25 years. When the time came for a change, it was offered to the UAS, if they could extract it from the school.

chat for years afterward, with recollections of a veteran Sea Fury (TF956) which stole the show—by accident.

Lt Cdr John Beattie had taken to the air only to find that when the time came to land, one undercarriage leg would not lock down. Society members suddenly found themselves hanging onto every word of the radio exchange between the pilot and his ground crew. Despite several vicious manoeuvres, the leg refused to engage,

ARDS SET TO TAKE OFF IN 'NINETIES

£2.5 million project planned

Story by Stephen Gordon

Ards Borough Council is poised to embark upon an ambitious new project for the 1990s — a £2.5 to £3 million 'Flight Centre' on the Portaferry Road, Newtownards, which it is hoped will attract over 100,000 visitors per year. The 'Flight Centre', which if it goes ahead will be the first of its kind in the UK, will be a science and technology centre focusing on all aspects relating to flight, from birds to the latest aircraft.

L&R Leisure's Flight Centre proposal was bigger, better and ultimately more expensive. The cost was too much for Ards council to bear.

leaving but one option. The visiting members looked on in horror as the aircraft slammed into the sea. Beattie had baled out, to be rescued by a Sea King helicopter. The Sea Fury, however, had sunk and was wrecked beyond repair.

The Society, meanwhile, had been focusing its efforts on an aircraft that could be rescued. Furthermore, there was money in hand to make it happen. A successful ballot in August, 1989 raised £2,500 for the heritage team and the Heritage Investment Project was bringing in over £100 per month. More than a year since the idea was first discussed, the Sea Hawk jet fighter could at last be saved. Society members had finally raised the money. Now they had to raise the roof. Fortunately, that could be done with the expertise of Heyn Group's Jim Walsh once more on hand. The Sea Hawk would have to exit the Shorts Apprentice Training School at Sydenham by way of the roof.

The Sea Hawk jet fighter could at last be saved.

Glass and steelworks were carefully removed, the wing-fold mechanisms freed, and the lift was on. It was October 14th, 1989. Riggers from Glovers' heavy-lift business trained the crane harness carefully, hauling their burden upward to within an inch of the trapdoor through the ceiling. Out came the jet fighter, to the relief of everyone at the scene.

From there, it was a relatively

easy exercise to tow the Sea Hawk across the Sydenham runway to store it with the Vampire and the big SD3-30 in the hangar of the Queen's University Air Squadron. The Society had added yet another aircraft to its collection, raised all the required funds voluntarily and once again had undertaken a significant engineering challenge as a group of amateur aircraft enthusiasts, under the

Members watched in horror as Sea Fury TF956 failed to lower its starboard undercarriage leg. The pilot bailed out of the aircraft moments later.

The Sea Hawk flies again: The delicate operation, helped once more by Jim Walsh, was a success and the UAS added another aircraft to the Sydenham hangar.

watchful eye of Jim Walsh.

Museum problems aside, the 1980s had ended on a high, and the '90s dawned with boyish optimism. Shorts had been saved from financial peril by Canada's Bombardier group. A whirlwind of change was in store for Northern Ireland in the decade that

followed, and also for the Ulster Aviation Society.

The highlight for many members was a group trip to Scotland in September, 1990, taking in various airports as well as the Leuchars air show. Members Roger Andrews and Bobby Bilsland kept faith for the spotters with an

A whirlwind of change was in store for Northern Ireland.

impressive list of observations for October's *Ulster Airmail*. However, the two-page report of the honorary secretary at the time, one Ray Burrows, was a jovial masterpiece of missing information about aircraft. He regaled his readers instead with anecdotes of a ferry that pitched and rolled, with passengers that nibbled and

The originator of low cost flying? Stephen Boyd negotiated cheap flights for members in the mighty 747.

tossed. There were reviews of meals and comely waitresses, general camaraderie, missing members and discreet suggestions of a too-good time. His full report on the air show itself was brief and to the point: "Lots of aircraft flying from 10 a.m. to 6 p.m." There's no question that aircraft spotting and photography were still major activities within the Society but, starting in the early 1990s, Stephen Boyd found a way to make the actual flying experience very popular. When he wasn't scrambling as editor to fill each issue of *Ulster Airmail*, he nursed his contacts at Aer Lingus. And that's how he introduced the Society to the mighty Boeing 747.

The carrier's scheduled flights to the USA and Canada from Dublin stopped regularly at Shannon to top up its jumbo jets with additional passengers. Stephen had a chat with the airline about the empty seats on the Dublin-Shannon leg of the journey, and they made a deal. Aer Lingus would offer those seats to Ulster Aviation Society members for minimal cost.

The first of those flights left Dublin Airport on June 22nd 1991, to touch down in Shannon after only 30 minutes in the air. There, the group was treated to tours of the ramp and air traffic control before returning to Dublin on a Boeing 737. The unique trip was hugely successful—so much so that several 747 arrangements were made in 1991; all of them sold out within just three days of being advertised.

"They went off absolutely brilliant," said Ray Burrows. "Fantastic! A flight in a jumbo for 20 quid? And then back again on a 737! I had the pleasure of actually going on one of those trips. It was superb, and that was the start of something pretty big then." But Stephen didn't stop there. Another chat led to Aer Lingus inviting Society members on low-cost trips from Belfast International to Dublin via Saab 340B. This was a case of the Society taking advantage of dead-legging: aircraft flying on positioning flights with normally no fee-paying passengers on board. Other flights were arranged with different airlines from Belfast to destinations in England. But that year wasn't all fine flights and sweet fares.

Late 1991 brought a surprise ultimatum from Queen's University Air Squadron: Haul your three aircraft out of the Sydenham hangar within the week.

With dim prospects on the museum front and a trio of complete aircraft in storage, the heritage project was caught in a potentially fatal position.

It was now that the meticulous research of chairman Cromie again paid off. He knew of a Second World War site once occupied by the US Army Air Force where now-vacant hangars still stood. He wrote

Control tower tours were a popular Society activity. Member and air traffic controller Ray Burrows wangled many visits including this one to the centre at Shannon.

Another coup for Stephen Boyd. Happy members board an Aer Lingus Saab 340 for another low-cost flying experience.

a letter to Marbur Properties, the firm that managed the former airfield. The response was positive: Your aeroplanes can have a home.

On October 16th, 1991 the dependable Heyn Group was called upon once more and the move began. Langford

He knew of a Second World War site where now-vacant hangars still stood.

Lodge was the destin-ation, a wartime airfield once home to the Lockheed Overseas Corporation and now the base of operations for Martin Baker in Northern Ireland. Unfortunately, the urgency of the move, and the aircraft's size meant that the SD3-30 prototype donated by Sir Philip Foreman had to be abandoned. (It ended up with the airport fire service and went from there to a scrap heap.)

The Society's frustrations with the flight centre flop and the abrupt boot from Sydenham had provided the leadership with the research and learning experience to better ponder the future.

The interest shown by the Ards Council in the mere possibility of an aviation museum forced the heritage group to seriously consider what such an attraction could look like. What displays would feature? What exhibits would be required? If not a proper museum now, what could an interim stage be? How would the finances figure into it all? The lengthy and difficult exercise prepared the Society for the next step in its story. In Langford Lodge it appeared they had found a long-term home. True, the buildings were more in need of restoration than the aircraft, but there was space and potential to develop, because Langford opened another door for the team.

Sydenham was an active airfield. That meant the public could not be brought on site

to view the heritage collection, such as it was in those days. Langford, however, was no longer regularly active and so the possibility of public visits could be entertained. It would not be easy. The site was home to an engineering firm which dealt with military hardware, meaning public visits would have to be prearranged and supervised. But they were visits nonetheless, and what's a museum without visitors?

"It was not five-star by any stretch of the imagination," said Ray, "but the hangar was 50 times larger than where we'd been in Newtownards. It meant a lot of cleaning and it had an asbestos roof, and we put a plastic sheet over the roof girders to stop any stuff from falling down. A lot of preparation work went into it, but we thought, 'Wow! What a place. We can bring our collection here, we can spread it out, we can see exactly what we have.'"

What they had was a growing collection—and a touch of variety. Shortly after moving into Langford Lodge, the team purchased a 1947 Bedford RAF refuelling bowser from the company. It marked the first departure from collecting aircraft and showed a good degree of foresight. The February 1992 *Ulster Airmail* noted that the bowser, with a bit of work, would look well displayed in conjunction with the Vampire or Sea Hawk. The experience of dealing with the flight centre proposal was paying off.

The Langford hangars had been neglected for years. But the Society was no longer at the mercy of Sydenham's airport rules, and the collection had space to grow.

NO SMOKING
WITHIN 50 FEET

The first ground vehicle purchased by the Society, this Bedford bowser, spoke volumes about the committee's long-term planning for a museum.

LANGFORD LODGE IN 1944, AMERICAN PLANES

*Aircraft fill every inch of space at Langford Lodge in 1944.
Thousands of American aircraft joined the European war via
Northern Ireland.*

LANGFORD LODGE IN 1944

On March 28th, the first planned visit to Langford Lodge took place—a group of 30 Ulster Aviation Society members. For many, the restricted access at Sydenham meant this was their first opportunity to view the collection their Society had been working hard to build for almost ten years.

For the majority of members, however, the main contact with the UAS was via *Ulster Airmail* and the monthly meetings; the latter did not disappoint with Bob Knights, who flew with the 'Dambusters' (617 Sqn, RAF) entertaining the gathering in April.

Shorts saw fit at the end of 1992 to donate the remaining pre-production SD3-30 to the Society.

Members continued to avail of flying opportunities, be it via the now regular Dublin to Shannon 747 experience or in the De Havilland Dove operated by RSJ Aviation, which was made available to members the evening before the Ulster Air Show on July 18th. The air show itself was once again adorned with the Ulster Aviation Society marquee, and visits to air shows nationwide continued.

The collection finally had a long-term home, members had more aviation experiences on offer than ever and acclaimed guests frequented the monthly meetings. The magazine was thriving, too, though the rarity of actual Society news appears in retrospect to be remarkable. There were pages—often filling half an issue or more—of airport movements and the arcane transcripts of airband radio chat. However, one could scan past editions of *Airmail* in vain looking for regular, detailed updates on heritage activities, though they were fast becoming the Society's most important enterprise.

In that regard, Shorts saw fit at the end of 1992 to donate the remaining pre-production SD3-30, G-BDBS, to the Society. That alone was bound to make the Society's 25th anniversary year an interesting one.

The aircraft remained in Shorts' facilities at Queen's Island while the heritage team prepared it for a fresh coat of paint and planned an ambitious logistical move to Langford Lodge. The operation would be no less complex than that which had scuppered G-BSBH, but now the members had time to plan ahead.

The Wildcat had already been moved by road with no difficulty. Maybe the SD3-30 could go on the same route. Once the volunteers removed the outer wings and tailplane, the aircraft could be transported by road as far as Nutts Corner. However, the narrow roads from that point to Langford Lodge meant that an inventive solution to the final leg of the trip would be needed.

The answer lay in a government scheme labelled Military Aid to the Civil Community. Ray Burrows and his colleagues figured an application for an RAF airlift would be a long shot but went ahead with it anyway. In fact, the commanding officer at RAF Aldergrove responded with enthusiasm at the prospect of such an unusual challenge.

There was, of course, a catch: The Society would have to insure the operation to the tune of £5 million. The high premium appeared to make the airlift idea the stuff of fantasy.

Ray had toxic visions of another monumental fund-raising effort, except for one possible option. With the help of an old friend who also had a friend, the proposal landed on the desk of insurance broker Archie McAvoy. Lowndes Insurance, he said, would take the case for an amazing

The Society's first book showed that the UAS was interested in all facets of aviation, not just heritage.

£200. As for Archie, he just wanted a photo of the operation for his brochure.

With all the paperwork in place and the RAF on board, the twin rotors of a massive Chinook transport helicopter cranked into action. On April 7th, G-BDBS touched down from its final flight to proudly join the Society's growing collection at Langford Lodge. It says something of the work ethic of the Society volunteers that the arrival of the Shorts SD3-30 at Langford could be challenged for highlight of the year.

In time, Society members and the public at large would know what obstacles these determined champions overcame. For the moment, though, they would read about Northern Ireland aviation in a form that was as ambitious as it was somewhat prosaic.

It came in the shape of a book. Three members (each one a committed spotter and/or photographer) compiled The *Ulster Aviation Handbook*, launched in June, 1993. While the heritage crew was occupied with establishing the Society at Langford Lodge, Stephen Boyd, Bobby Bilsland and Jack Woods had been plugging away at research and writing. Their publication, with input as well from Ernie Cromie and Ray Burrows, charted all aspects of aviation in Northern Ireland as of 1993. Readers could learn about the country's airfields large and small, active and disused. Sections were dedicated to commercial and

private flying instruction and aerospace manufacturing. At a time when the recent focus on heritage could have been seen as polarising, the publication of *The Ulster Aviation Handbook* was refreshing to all members.

"That was one of the things about the Society," said Stephen Boyd. "It encompassed a lot of different types of aviation interest."

Between the spotters and the heritage group, it could be easy to lose sight of the point that most Society members seemed eager to learn all kinds of things about aviation. Within the *Airmail* there was an increasing number of regular, well-researched historical articles appearing from time to time, usually courtesy of the heritage faction. Added to that, at Society functions— especially the monthly membership meetings—guest speakers covered a wide variety of subjects. Some guests were worth books of their own. In September of 1993, Ulster's last five Battle of Britain veterans were Society dinner guests. Graham Mehaffy put that into perspective 25 years later: "We had some absolutely fantastic speakers," Graham recalled. "And you know, those guys are all gone now—the World War Two guys."

Among the speakers of that period he mentioned Bill Reid, Victoria Cross bomber pilot who continued his mission though seriously wounded by a flak burst. There was Ken McKenzie, who downed

After a fresh coat of paint, G-BDBS is readied for the road. The Society was extremely lucky to get a second shot at owning an SD3-30.

"We had some absolutely fantastic speakers."

a Messerschmitt fighter by flicking its tail with his own wing. Charlie McIlroy, a Red Arrows pilot, gave a fascinating presentation. Sir Michael Bishop, chairman of British Midland, showed up and shortly afterwards British Airways, not to be outdone, sent along Senior First Officer Mick Crossey. Along with his talk, Mick brought two ballot prizes: a pair of tickets to London

From the passengers' perspective: Aviation enthusiasts don't consider the wing and engine nacelles an obstruction! The Douglas DC-8 would soon be seen no more; the aircraft retired from Translift's fleet in 1995.

Heathrow, complete with a visit to the simulators at the BA training centre. Spitfire pilot Raymond Baxter, presenter of BBC television's *Tomorrow's World*, packed the room. And the list goes on, an impressive tally during that period in the late 1980s and through the '90s. Ray Burrows called these the Society's halcyon days.

Through it all, various trips continued. For example, 44 members bit into an offer of a flight on a Translift DC-8, considered by many to be an 'endangered species' in 1993. And twice that number headed off to Manchester on a day trip via a Jersey European BAe 146-200 on October 2nd. On the ground in Manchester, some availed of the opportunity to see the local science museum. Others visited the city centre shops while a few diehards loitered at the terminal, noting every aircraft registration in sight. "Going on pleasure trips was absolutely superb and a lot of the members supported us," said Ray, whose air traffic control duties prevented him from going along regularly. "We were never short of members who wanted to go flying."

Visits to Dublin Airport and the Leuchars air show—an annual trip which consumed a serious amount of organising and associated frustration—continued as normal, of course, and a team of UAS volunteers waved the flag once again at the Ulster Air Show at Newtownards. Unfortunately, at the same location just two

months later, the Ulster Flying Club advised the Society it would have to move its monthly gatherings from the premises. The meeting room was being sold to a restaurateur.

Surprised (but accustomed by now to moving around), the Society's management committee quickly set about finding alternative accommodation and by year's end Jack Woods had secured regular space in the building of the Church of Ireland Young Men's Society (CIYMS) in east Belfast. With a few exceptions, meetings have been held there ever since.

Meanwhile the heritage team had been quickly but quietly working on yet another landmark acquisition—one which presented itself rapidly, with little time for the full Society leadership to assess in detail. However, with some help from Secretary of State Sir Patrick Mayhew, the details were finalised and at the AGM in March, 1994, the announcement was made: The Society had purchased an RAF Blackburn Buccaneer. The distinctive strike aircraft (XV361) was one of the last flying examples of the type, and that brought a special advantage: The jet would be delivered under its own steam, so to speak.

Up to then, the major disadvantage of being island-based was the cost to the UAS of transport, compared to the costs facing heritage groups based in GB.

The Heyn Group were back on hand to help with the first leg of the journey by G-BDBS from Sydenham to Nutts Corner.

XV361 touched down at Aldergrove on April 5th after departing from Lossiemouth and completing a farewell tour of ex-Buccaneer bases.

But how to get the new exhibit to Langford Lodge: That was the question.

The aircraft was far too heavy for a Chinook airlift, so plans were researched to move it by road. Preparations were under way when a third option occurred to the team: Why not fly the 'Bucc' to Langford Lodge? There were no control tower facilities, but the runway was long enough. It would just be a case of zoom up, zoom down and park the beast.

British Airways were once again generous with their support

> **XV361 departed Aldergrove and touched down at Langford a mere 92 seconds later.**

of the Society and flew the two-man crew at no cost from Inverness to Aldergrove to prepare the aircraft for its final flight. On the cold, dull morning of April 18th, under the command of RAF S/Ldr Martin Hopkins and F/Lt John Parsons, XV361 departed Aldergrove and touched down at Langford a mere 92 seconds later. It was believed to be the shortest-duration trip a Buccaneer had ever flown. Another historic aircraft had been added to what was taking shape as the biggest collection of heritage aircraft on the island of Ireland. It's worth considering the significance of BA ferrying the crew free of charge. The Ulster Aviation Society was clearly held in high regard by the airline giant, as it was also by Air

Lingus. The flag carrier of the Republic was again offering members low-cost flights from Dublin to Shannon, this time in its ageing Boeing 707 fleet, throughout 1994. In fact, the Society's status had grown among the aircraft industry far and wide, evidenced when Ken Ellis, distinguished editor of *FlyPast* magazine, arrived as a guest speaker and left as a Society member.

As 1994 came to a close, it had become clear that the reputation of the UAS had been enhanced largely due to the expanding collection and the attendant publicity.

Chairman Cromie made note of it in his annual report: "I find it difficult to remember a year in which the Society's

XV361 served with both the Fleet Air Arm and Royal Air Force before retirement placed it in the hands of the Ulster Aviation Society.

name was so often in the news for one reason or another."

Not everything reached the eyes of the news media. Regrettably, dissension still bubbled within the management committee. At one point, a key member resigned.

"The attitude of some committee members appeared to be at worst hostile and at best completely disinterested and unsupportive towards our heritage projects," said his resignation letter.

That sparked a special committee discussion in June, with knuckles rapped and apologies tendered. The discouraged member withdrew his resignation.

Complaints from members about a lack of heritage information led also to a decision that the *Airmail* start providing regular reports on Society activities. Such events seem to have marked a positive turning point between the leadership and the members in terms of accountability and transparency.

Now it was time for the Ulster Aviation Society to sell itself to the public at every opportunity, to build support and trust with local councillors and businesses. To this end the committee planned the Society's first open day at Langford Lodge. It was billed as a 'thank you' to members and took the form of a fly-in and barbecue on August 27th.

Buccaneer on approach. The 92-second flight from Aldergrove to Langford is believed to be a world record for brevity.

Thirteen fixed-wing aircraft and three helicopters were directed by Allan Boston, who had the honour of handling the Langford traffic on the site's busiest day since the Second World War. First on the ground was Chipmunk

Flying an active military aircraft into the Society's collection demonstrated an ambition out of all proportion to the group's voluntary, non-commercial status.

Visitors get up close and personal with a Westland Wessex during a Society fly-in. The events would grow exponentially year on year.

G-AOTR piloted by Mike Woodgate, a Society supporter from its earliest days.

The year 1995 would not see the same focus on aircraft acquisition as those that preceded, but the Society still faced its fair share of opportunities and challenges. Indeed, the 50th anniversary of Victory in Europe (VE Day) prompted groups throughout the whole country to mark the occasion. Two subtle factors likely motivated their initiatives: one, the aircrews who fought in the war were not getting any

The VE Day anniversary would be a theme for a fly-in

younger, and the anniversary would provide the public with a chance to thank them for their contributions. Secondly, many of the organising societies—and the UAS was a prime example—were run by the next generation. They were often "baby boomers", strongly influenced by their parents' wartime experiences. Aviation enthusiasts in particular, from their days as awestruck youngsters full of national pride, had chosen as their heroes those brave and daring young men who flew during the war and who also played such a major part in

the U.K.'s impressive postwar aviation developments.

Here in Northern Ireland, VE Day could be an occasion for celebration, and the Society seized upon the idea. It would organise an open day for the public, on a much grander scale than the 1994 event. How about a fly-in, Society displays, live music, a barbecue and maybe—just maybe—the appearance of an old warbird?

On May 20th, an unprecedented crowd of 5000 gathered at Langford Lodge. Over the course of the day, 38 light aircraft and helicopters flew into the airfield. Visitors enjoyed a vintage cars display, model flying, a planetarium, the Queen's University Air Squadron and the Army Air Corps Blue Eagles helicopter demonstration team. The heritage collection, small but growing, impressed the guests, many of whom had never before heard of the Ulster Aviation Society. The warbird? This was indeed the highlight of the day. At 4.30pm, the first B-17 Flying Fortress to visit Langford Lodge since the end of the

Second World War rolled down the runway and gently took to the skies. The flypast of "Sally B", herself 50 years old, was at once a tribute to the thousands who had worked at Langford Lodge in more perilous times, and an appreciation of the hard work of the Society's volunteeers. Some 80 members stepped up to make a success of an event never undertaken before by the Society on such a scale.

To their credit, the organisers elected not to profit from the celebration, feeling to do so

The pamphlet created for the open day explained the context of the event for those new to the UAS.

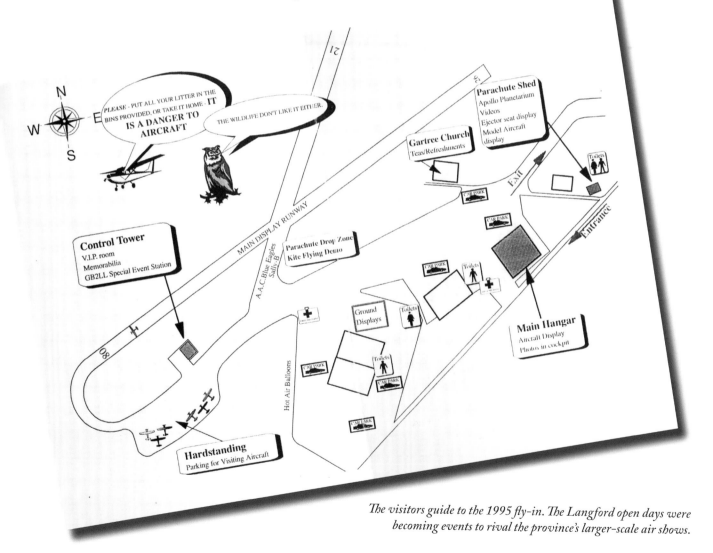

The visitors guide to the 1995 fly-in. The Langford open days were becoming events to rival the province's larger-scale air shows.

The B-17 made a poignant return to Langford in the form of 'Sally B'. The last time B-17s visited the station they were on their way to war.

The committee organised a spectacular event: aircraft flypasts, vintage jeeps, ground displays, tours and a hot air balloon. Everywhere a visitor turned there was something new to see.

would degrade the intended commemoration of the end of the war. Fortunately, generous donors and sponsors bore much of the cost.

The remainder of 1995 passed relatively quietly given the flurry of activity the Society had been used to in years past, though the committee continued to plan amazing activities for the Society community. The September meeting was a particular highlight, when at the Airport Hotel at Aldergrove the members sat in contemplative quiet as five ex-882 Squadron FAA pilots, all of whom had flown the Society's Wildcat, regaled their audience with

> **The monthly talks of '96 could be considered the best ever.**

tales of carriers, camaraderie and combat. On October 25th, 30 members jumped at the opportunity to experience a short flight for only £7 in a Gill Airways ATR 72—a competitor with Bombardier's own Dash 8. It would be the first of many successful trips to be arranged by committee member and prolific aviation writer Guy Warner.

As the year wound down, the Ulster Aviation Society returned to its roots when a gaggle of spotters gathered on November 30th at the gates of Aldergrove in the hopes of catching a glimpse of VC-25A, a converted Boeing 747. It's better known by its call

sign, Air Force One. It landed with U.S. President Bill Clinton aboard. He was about to herald, hopefully, a new era of peace for Northern Ireland. The very fact that spotters and photographers could ply their hobbies at or near Aldergrove, with limited security precautions, was perhaps a cautious sign of progress as far as aviation enthusiasts were concerned.

Six months after the president's visit, UAS members headed south of the border for another "presidential" appearance as the huge American aircraft carrier USS *John F. Kennedy* dropped anchor off Dublin. Unfortunately, repair work on

U.S. President Bill Clinton's arrival in Belfast in 1995 signalled a new era for the peace process. Society member Jack Woods snapped this shot of Air Force One as it landed at Aldergrove.

the damaged hull prevented UAS members from boarding.

While the years that followed didn't feature the aircraft acquisitions or exciting events of the early 90s, the committee work behind the scenes would have a far greater impact on Society operations in the long term. First announced at the AGM in 1996, the intent was—with the agreement of the membership—to apply for charitable status. Money was the main consideration. It seemed that as a charity, new revenue streams would open for the Society, with no particular disadvantage to be seen. The onus would be on the organisation to put the new

funds towards the education of the Northern Irish public.

'Education' was undefined, thus the widest meaning was possible. The Society was already doing this at air shows and with tours of the heritage collection, for example. Charitable status seemed like a winning proposition to all concerned. In fact, the upcoming May bank holiday would see the biggest visiting group ever, 150, to Langford Lodge. Intended to be a brief pause on an all-Ireland coach tour, the group stayed all day, such was the impression the collection and its curators made. The task now was to re-word the constitution appropriately, get approval from Inland

The USS John F. Kennedy *anchored in Dublin Bay in 1996 to enthusiastic crowds. A limited number of visitors were ferried out to the carrier, but Society members were able to view a static display at Dublin Airport of aircraft from the ship.*

Members gather in front of Hunting Cargo's Merchantman 'Superb.' The accompanying video was a first for the Society.

Revenue and endorsement by the membership.

The monthly talks of '96 could be considered the best ever. In March, the Red Arrows team manager, Red Ten, was greeted by 112 eager faces. They then witnessed the arrival of Bell Jet Ranger pilot Nick Bather who dropped in by helicopter at CIYMS. Sixty-eight lucky members and friends enjoyed aerial tours of east Belfast in the Ranger, though some locals were less than amused by the noise.

Flights for members continued to be a hugely successful endeavour, and now the Society decided to use its airline contacts to diversify in an altogether new field: film. On September 29th a team from the Ulster Aviation Society were privileged to witness, and film for posterity, the last operational flight from Belfast International to Coventry of Vickers Merchantman G-APEP, the freighter variant of the Vanguard. Hunting

Vickers Merchantman: Belfast to Coventry *was the Society's first foray into video production. The Merchantman was a Vanguard converted to an all-cargo layout; only nine were so modified.*

Cargo Airlines welcomed the Society for this historic flight. The camera work was followed by several months of scripting, voice-over recording and editing. What emerged was a very professional production, impressive for what continued to be a voluntary, self-funding organisation fuelled almost entirely by enthusiasm.

For a group with a strong emphasis on heritage, there were some in the Society with their eyes on the future. To this end, Michael McBurney founded the UAS website on March 9th 1997 with the rather fussy address: http://www.niweb.com.dnet/dnterAzQ/. By July the "hit" counter had struck 600, with visitors from 26 countries as far afield as Malaysia, South Africa and New Zealand.

Closer to home, the Society's cross-border relations were also healthy. The Aviation Society of Ireland invited the Ulster Aviation Society to join them for a Dublin to Shannon flight on a Lockheed Tristar. Even faster than a Tristar but here on terra firma, an RAF wing commander blasted through the sound barrier on October 15th, 1997. Andy Green spurred his twin Spey engines in a Thrust SSC to 763 mph to break the world's land speed record. Andy, by the way, had won his jet fighter wings in a Phantom—a favourite type which joined in the Society's collection in 2015 and powered, of course, by twin Spey engines.

Not quite as glamorous, but back home in Northern Ireland the members of the Ulster Aviation Society were breaking new ground of their own. At a special general meeting in November, 1997 the lingering notion of becoming a charity was resolved. The committee assured the members that day-to-day Society activities wouldn't be affected by the new status and a large majority agreed the proposal. Shortly afterward, the Society was accepted as an educational charity by Inland Revenue. It was another landmark in the Society's history, and one which would have lasting implications.

AIR TRAFFIC CONTROL

VISITS 1986

Regarding our recent talk on a proposed visit to Air Traffic Control at Aldergrove, we are now taking details to help us facilitate this matter.

Please fill in the form below and return to:
Ulster Aviation Society, Hon. Secretary, 20 Carrowreagh Gardens, Dundonald, BT16 0TW.

Name ....
Address ....

Date of Birth ....
Car Registration No. .... Colour/Make ....

Preferred time of visit
- ☐ SATURDAY MORNING
- ☐ SATURDAY AFTERNOON
- ☐ SUNDAY MORNING
- ☐ SUNDAY AFTERNOON

PLEASE TICK ONE ONLY

FURTHER INFORMATION AND DATES WILL BE AVAILABLE AT A LATER TIME

DUBLIN AIRPORT VISIT
(Les Thompson Memorial Trip)

Shake off those winter blues and join the UAS visit to DUBLIN AIRPORT on SATURDAY 21ST MARCH 1987.
Leaving Glengall Street   Ulsterbus Depot 08.00
                            QUB Student Union 08.05 (easier & safer parking)
                            Blaris/M1 Roundabout 08.15
                            Newry/Rockmount 08.55

The visit is timed to coincide with the Ireland v. France rugby match. It is hoped to include a Ramp Tour and a visit to the Irish Aviation Museum Store to see their Provost and Vampire. Estimated arrival in Belfast 18.00 hrs.
Cost £7.00 per person. Interested? — Complete the form below and post details to **Stephen Boyd, 7 Malone View Road, Belfast BT9 5PH.**

Name ....
Address ....
Telephone (Day) .... (Night) ....
Please reserve .... seats at £7.00 each. I/We wish to board at Glengali St/QUB/Blaris/Newry.
I enclose PO/Cheque for £............ payable to UAS and I understand that refunds will be given only if my seats can be re-allocated.
Signed ....

DUBLIN TRIP

Including visit to Aer Lingus Maintenance Facility and Airport Ramp will now take place on Saturday 13th October 1984.

Minibus will route Bangor/Upper Newtownards Road/Castlereagh Road/ Maysfield Leisure Centre/Donegall Road (M1)/Dublin.

. approximate cost for Minibus is £7.00 per person.

If interested please complete and return this form immediately to:
MARTIN WALLACE, 69 BALLYSALLAGH ROAD, CLANDEBOYE, BANGOR.

Times for pick up points will be notified later.

NAME ....
TELEPHONE NUMBER ....
PICK-UP POINT ....

DUBLIN AIRPORT TRIP

10th MAY 1986

Following the success of previous visits to Dublin, a trip has been arranged for Saturday, 10th May. Included in our visit will be a morning ramp tour, Aer Lingus maintenance facilities with the afternoon free.
Coach will be departing from:
GREAT VICTORIA STREET ULSTERBUS DEPOT at 08.00
Pick up M1 BLARIS ROUNDABOUT 08.10
On the return journey we will depart Dublin at approx. 17.45 and arrive back in Belfast at 20.00.

The cost for the trip is £6.50. Deposit £2.50 with the balance due by 30th April 1986
Cheques payable to UAS via
STEPHEN BOYD, 7 MALONE VIEW ROAD, BELFAST BT9 5PH

Name ....
Address ....
Day Time Phone No. .... Evening Phone Number ....
Membership No .... Cheque/PO £ ....
Boarding Belfast ☐   Blaris ☐

PLEASE TICK AS NECESSARY

0790
ULSTER
AVIATION
SOCIETY

0790   ULSTER
AV   TION SOCIETY
Grand Draw
(Proceeds in aid of Wildcat Restoration Fund)

Draw will be held at Aldergrove Air Show, 17th August 1985

1st Prize: TWO RETURN SHUTTLE TICKETS
TO LONDON (Donated by British Airways)
2nd Prize: £50.00
3rd Prize: Flight in Cherokee Aircraft
4th Prize: Selection of Bushmills Whiskey
Plus 10 Consolation Prizes of £5 Each.

TICKETS ....£1.00 each

This Ticket is a receipt for a voluntary subscription and is given and accepted as such

---

ADVANCED SALES

ULSTER AIR SHOW

SATURDAY 18TH JULY 1992

ULSTER AVIATION SOCIETY
presents
White Lightning Disco

****

in
THE ULSTER FLYING CLUB, NEWTOWNARDS
on
FRIDAY, 30th November

****

DANCING 8 - LATE
Ticket £4.00 Including Supper (Chicken or Scampi in Basket)

THE ULSTER AVIATION SOCIETY
in association with

LOGANAIR
Scotland's Airline

present a
PLEASURE FLIGHT IN A TWIN OTTER

ON SUNDAY 22ND JUNE 1986
DEPARTING FROM BELFAST HARBOUR AIRPORT

Please report at Belfast Harbour Airport at .... p.m.

FLIGHT TO BE OF APPROXIMATELY 30 MINUTES DURATION

IRISH AIR SPECTACULAR

SUNDAY 17TH AUGUST 1986
at
BALDONNEL

There will be a Day Trip by Ulsterbus Coach departing
GLENGALL STREET at 07.45
BLARIS/M1 ROUNDABOUT at 07.55
ROCKMOUNT ESSO STATION at 08.30

The Coach will leave Baldonnel at 18.00 hrs. and will arrive in Belfast at approximately 20.45 hrs.
COST £7.00

Please forward Cheque/Postal Order made payable to Ulster Aviation Society for £7.00
Name ....
Address ....
Tel No. (Day) .... (Evening) ....
OPEN TO NON MEMBERS
ABOVE TIMES HOLD TRUE UNLESS NOTIFIED TO THE CONTRARY

Please forward to: STEPHEN BOYD, 7 MALONE VIEW ROAD, BELFAST BT9 5PH. Tel. 616659

ULSTER AVIATION SOCIETY
in conjunction with
AIR ATLANTIQUE

invite you to join them for a pleasure flight in the famous

DOUGLAS DC-3 DAKOTA

ON SATURDAY 27TH SEPTEMBER 1986
Departing from Aldergrove Airport - Check in Gate 9
Please report at ....

Certificate of Flight
THIS IS TO CERTIFY THAT
COLIN BOYD
has flown on a

DOUGLAS DC-3 DAKOTA
which departed from Belfast International Airport
on Saturday 27th September 1986
Signed ....
(On behalf of Ulster Aviation Society)

The Ulster Aviation Society
present a
Disco

IN THE ULSTER FLYING CLUB
NEWTOWNARDS
28th on
FRIDAY, 21st NOVEMBER, 1986
7.30pm to Late
Admission by Ticket only   £2.00

№ 2   № 862   ULSTER
                        AVIATION SOCIETY
Ulster
Aviation Society
GRAND DRAW
(Proceeds in aid of Wildcat Restoration Fund)

Draw will be held at end of Show at Society Marquee

1st Prize: TWO SHUTTLE TICKETS
TO LONDON (Donated by British Airways)
2nd Prize: £25.00
3rd Prize: £10.00

TICKETS — / — / — £1.00 each
This ticket is a receipt for a voluntary subscription concept entry

---

LAST CHANCE - FOR AER LINGUS BOEING 747 FLIGHT

SHANNON Airport -day trip by Flexibus and AER LINGUS Boeing 747-100/737-500.
On Saturdays 17 September and 1 October. From Glengall Street 07 00 by FLEXIBUS
to Dublin EI133, a Boeing 747-100 service to Shannon with anticipated 5 hours
in Shannon before return on EI092 Boeing 737-500 to Dublin and bus to Belfast at
20 00 approx. This will be the last opportunity to fly on an EI 747 before their
forthcoming retirement this winter. Ramp tour and ATC requested at Shannon.

Send 1) SELF STAMPED ADDRESSED ENVELOPE to Stephen Boyd
15 Broomhill Park Belfast BT9 5JB with CHEQUE payable to
ULSTER AVIATION SOCIETY for £ 40 -per person.

Names .... Address ....
Telephone Day .... Evening ....
Booking non refundable after 9 September.

LEUCHARS 1995

ONCE AGAIN IT IS THAT TIME OF YEAR AND ONCE AGAIN WE ARE PLANNING FOR
YET ANOTHER SUPERB WEEKEND AT St.ANDREWS.   THE AIRSHOW HANDOUT
THIS YEAR BOASTS SOME NINE NATO AIRFORCES PARTICIPATING AND PROMOTES
IT AS THE BEST SHOW EVER!   BESIDES THE AIRSHOW CAN YOU AFFORD TO
MISS THE CRACK?   BEST NEWS OF ALL - THE PRICE HAS GONE DOWN........ TO
AROUND £85.00 !

There are many things to consider when gathering a collection of historic aircraft. Which aircraft do you select? How best to display them? To what degree do you restore the aircraft, or allow the wear and tear of an active life to show through?

Perhaps most significant, though: How do you move an aeroplane to your museum site? The Ulster Aviation Society has been faced with this conundrum on many occasions over the years and has responded with some ambitious, clever and often unique approaches.

The Wildcat

The first, of course, was the recovery of Wildcat JV482 in the mid-eighties. The engine and propeller had been lifted successfully from the silt of Portmore Lough in late 1983, leaving the main body of the aircraft for another day. An initial attempt had been made at the close of 1983 but the silt-filled fuselage proved too heavy. That left the recovery team to toil tirelessly to lighten the wreck and lift it to the lough's surface. Such a project allowed the ingenuity of the members to shine.

"To get out to the Wildcat…Raymond McMullen constructed a barge made up on 100-gallon drums, welded together then fitted an outboard motor to it. We used to motor out to the wreck on Raymond's contraption!" laughed Ray Burrows.

The team tried various ways to raise the hulk, finally settling on more large drums. They rolled them under the wings and filled them with air from the divers' tanks. The resulting buoyancy gradually eased the fighter to just above the lough surface. The team then drained the water from the structure and cleaned the silt from its crevices.

The UAS team assisted by Belfast Sub Aqua Club and Jim Walsh installed pontoons and shovelled silt. Slowly the aircraft emerged from the shallows.

On the late afternoon of Monday, April 30th, 1984, Lynx XZ665 touched down on the shore of Portmore Lough and the stalwart sergeants, Shailer and Wood, made their last inspections, satisfying themselves that the wreck could be lifted safely. A crowd began to form on the shore in quiet anticipation of the events to come. Content with the lifting arrangements, the pilots returned to their craft. The ground team made for shore, leaving John Hewitt and Jim Walsh behind to do the honours. For a final time, the air filled with the ferocious roar of jet-powered helicopter blades as XZ665 manoeuvred slowly into position. Caught in the downwash, it was all Messrs Hewitt and Walsh could do to make the connection between helicopter and aeroplane before rowing with difficulty away from the furore. Time seemed to slow for the onlookers, as the daredevil pilots tested the load. Inch by inch, the Lynx climbed, the tail of JV842 tilting up as well. The pilots paused, allowing their charge to stabilise against the breeze. Then, with a fierce rise in pitch, engines screamed, rotors strained and the Wildcat soared free of the shallows for the first time in 40 years. The Society members on shore breathed a sigh of relief. The hard part was over. JV842 touched down on the shore and soon a crowd amassed, members and locals and media alike. The restoration began immediately; the aircraft could not be moved further until the wings were stowed. Volunteers spent a full day lifting the airframe off the ground to enable the undercarriage to be lowered and freeing the wing-hinge joints, until eventually the wings swung gracefully into their folded positions. The unrelenting support of George Heyn shone once again as his chartered lorry arrived to collect the relic for transport to Castlereagh College, the first of its several homes on dry land.

The SD3-30

Week after week through the early spring of 1993, the forlorn fuselage of the Shorts SD3-30 nestled next to the yard of a timber-shredding firm at Nutts Corner. Piles of nearby wood chips were bound for the floors of chicken coops; the aircraft was bound for modest glory. The wings and tail of G-BDBS would go by road to Langford Lodge. The fuselage would join them by air, courtesy of a Royal Air Force helicopter, as soon as a senior RAF officer was satisfied that the Society had the proper insurance. Ray Burrows recalled the discussion.

"He said, 'What sort of insurance do you have?' I said, 'Don't be silly! We don't have insurance. We're a voluntary organisation. If somebody hits somebody's finger with a hammer, you just kiss it better and tell him to go home!'"

The chairman walks on water: the sunken fighter was too heavily laden with silt to be lifted easily. It was dangerous too, as Ernie who slipped into the mire would testify.

Back on dry land after 40 years in the lake, the Wildcat represented a turning point for the Ulster Aviation Society.

With a rush of rotors, the Wildcat soared free of the lake. John Hewitt and Jim Walsh (in the row boat) were caught in the dangerous downwash.

3.1 How to Move an Aeroplane

A kiss would not work this time. But a bit of scrounging netted adequate coverage for only £200, from lift off to landing. On April 7th, the defining thud of twin rotors marked the Chinook's arrival over the treetops. But G-BDBS was not about to go quietly. As the mighty aircraft manoeuvred into place, the inexperience of the Society team met with the realities of powerful fan action from the huge rotors above.

"We hadn't put any chocks under the wheels," Ray explained. "So when the Chinook came up over the '330, the '330 rolled off down the tarmac by itself with no one in it."

It was a sight not easily forgotten: A frenzied bunch of men scrambling after half a large aeroplane as it limped away in a desperate run for freedom.

The Chinook crew above, doubtless choking back their laughter, was waved off while the ground team caught up with their quarry and chocked its wheels. But G-BDBS had another stunt to perform, and Ray had a clue what it might be.

He warned the RAF handlers that their rigging might not be complete. There were slings on the wing stubs and rear fuselage, but none on the nose. The RAF wasn't worried, and the SD3-30 left solid ground one last time, hauled by a straps-and-cable arrangement extending about 70 feet beneath the Chinook.
The SD3-30 design, it should be said, was such that the wide, flat bottom of the fuselage augmented to a small degree the slight lift capabilities of the remaining wing stubs. Perhaps they all sensed a chance to fly once more, like the old days.

"Away it went, with the (untethered) nose dropping about 15-20 degrees," said Ray. "The Chinook was getting faster, and when it reached maybe 40-50 miles an hour, the '330 started to fly itself."
Up went the nose of the '330, to plus 15 degrees or so, enough to stall and drop smartly, then rise yet again with the increasing speed of the helicopter above. The nose of the suspended '330' bounced at a higher angle each time, to the horror of the men watching below.

"Well, I can tell you we were almost wetting our pants," said Ray. "We reckoned the cable would be cut and the 330 dropped in a field, and that would be the end of it."

Luckily, an alert crew member aboard the Chinook spotted the problem. He told the pilot, the helicopter slowed down, the bouncing

The team inspect the lifting provisions. The arrival of the Chinook is imminent.

stopped and the awkward assemblage headed for Langford Lodge at a very poky pace.

"And when he set her down," said Ray, "he did it very, very gently."

The Phantom
Airlifts of ageing aircraft may look spectacular, but no less ambitious was 2015's recovery of Phantom FG.1 XT864. Having served with both the Royal Air Force and the Royal Navy, XT864 spent its twilight years as a gate guardian at RAF Leuchars. But as the force vacated the base to make way for the Army, the Phantom's use had come to a sad end. Still, there was hope for the aircraft: The Ulster Aviation Society submitted a bid for it. XT864 had, after all, spent short bits of its active life in the hands of 23 Maintenance Unit at RAF Aldergrove. The Society's bid was successful, so all that remained was to repatriate the fighter from Scotland to Northern Ireland.

If only it was that easy.

Unlike most of its contemporaries the Phantom was never designed for wing removal. The engines could be—and were—extracted from the aft of the aircraft negating the need to break the wing, which is one solid structure, 38 feet from tip to tip.

"Whenever we're looking at the wing and fuselage attachment, it's not straightforward. There are flanges, there are all sorts of things

After a false start when the aircraft rolled down the runway, the SD3-30 finally took to the air in the Society's most spectacular operation to date.

riveted and bolted," team leader Ray Burrows lamented at the time.

But if the beast of an aeroplane was to make it by ferry across the Irish Sea, the recovery team from the Ulster Aviation Society would first have to separate wing from fuselage.

Eleven trips were made to Leuchars, where the team worked tirelessly. Decades of active service, followed by the time and weather effects of 25 years outdoors compounded an already daunting task. This was relentless labour, taking strength and imagination to deal with challenges major and minor.

"A lot of the bolts are pretty well corroded, there's five or six layers of paint over them," said Ray. "It's an absolute nightmare."

The volunteers employed such delicate tools as sledge hammers, angle grinders and acetylene torches until, on June 13th, the wing popped and was parted from the fuselage. Bulky though it was, the fuselage was reasonably narrow and sat quite comfortably on a low loader with a long wheelbase. The wing was the issue: too tall to stand on edge, too wide to lie flat on a low loader. David Jackson and Geoff Muldrew got to work, designing and building a large, steel frame to secure the wing at an oblique angle. This cradle would hold the wing in the optimal position to avoid the various obstructions on the road home. The cradle itself, not a small item, had to be completed on site at Leuchars, the finishing

The bespoke cradle had to be modified on site for the perfect fit, but to the credit of Geoff Muldrew and David Jackson, it worked flawlessly.

It took several working trips to Leuchars to separate the wing but on June 13th, 2015 the team succeeded.

touches being made as the wing was lowered into place. On November 23rd, the wing arrived at the Society's Maze/Long Kesh location, five months behind the fuselage. It was eased slowly under the raised hangar door with easily two inches to spare. The planners had done their work well.

Years of experience had given the Ulster Aviation Society team the foresight to plan the operation in detail before ever starting. The design, construction and success of the wing cradle alone spoke volumes about a team, once amateurs, who had earned the accolade of the first civilian organisation in the UK to separate a Phantom wing. And the team's transport frame had attracted attention as well. It returned to Leuchars to assist the Cold War Jets Collection. The museum, at Bruntingthorpe, had purchased XT864's sister Phantom, XV582 *Black Mike*. The Society advised them on the dismantling process and also loaned the tools required to do a similar task in order to get their new acquisition to Bruntingthorpe.

These are just three stories among many from an exciting Society history of aircraft recovery. Each acquisition has had its story to tell—some of them just as amazing. The Sea Hawk was lifted through the roof of Shorts Apprentice Training School. The Buccaneer arrived by way of a 92-second flight to Langford Lodge. The Fairey Gannet turned up in nine major pieces and box after box of loose parts. The Canberra, the Society's largest aircraft, had to be dismantled piece by piece, transported from England and—on an even larger scale than the Phantom—reassembled in our hangar by an enthusiastic team which had to learn by doing. Each aircraft recovery has been an adventure in itself, each one adding to the experience of what has become one of the most respected aviation heritage groups in the United Kingdom.

After a year of journeys to Scotland the last remaining component of the Phantom finally hits the road.

CHAPTER FOUR

Almost without realising it, the Ulster Aviation Society was learning how to be a museum, and much more besides. Only 30 years earlier, key decisions by a tiny knot of plane spotters had involved the cost of new binoculars, schedules for the next bus to Aldergrove and the accuracy of their notes about the latest airport arrivals.

Now, at the close of 1997, the Society had several impressive aeroplanes of its own, leasing a large hangar with an historic aviation background. Collectively, members had travelled thousands of miles by air to events and museums. They had published hundreds of issues of a specialised aviation magazine, raised respectable amounts of money, recovered aircraft and assembled some impressive artefacts. As well, they were well organised and led, with monthly meetings and a formal constitution. They were catering to all sorts of aviation interests and had actually begun to prepare a museum of flight for the people of Northern Ireland.

Sadly, not all felt that the Society supported their individual interests. As with any dynamic organisation, some members felt somewhat disenfranchised. On December 16th, 1997, almost 29 years to the day since the formation of the Society, Stephen Boyd and Jack Woods hosted the first informal meeting of those members. They met, as in the old days, at Aldergrove—the

Sadly, not all felt that the Society supported their individual interests.

Photographers Stephen Boyd and Jack Woods were frustrated that Society meetings focused on guest speakers and not the members, so they established their own series of members' meetings.

Society's birthplace. It was not a palace revolt but just a spotters' gathering, with no talk of history, wreckology or museums. As Stephen Boyd explained in a recent interview, the evolving thrust of the Society at the time seemed to be concentrated on exploring the past.

"It was all about museums, it was all about aircraft restoration, and that really wasn't what a lot of us felt," he said. "Our interests were very much more in the present day....The two things that those meetings offered was, one, you were at an airport and the second part was there

Fred Jennings has been the keeper of the UAS library for over 10 years. As the Society grew, so did the range of activities and responsibilities undertaken by its members.

were no guest speakers; these were members' own slides of where you've been, what you've done…much less formal."

Eventually, that small subset (25 at most) drifted away from a formal connection with the Ulster Aviation Society. Those enthusiasts—spotters and photographers, mainly—continue to meet each month to socialise and view each others' photos. Several are Society members.

Past chairman Ernie Cromie shrugs it off, pleased that they

satisfy their aviation interests, but noting that the Society itself doesn't discourage spotters or photographers—indeed, encourages them: "We're a broad church."

Within that community, an increasing number of opportunities had arisen for those able to see that the organisation could be more than just a heritage centre. It has served, for example, as a resource for members to develop new skills in all sorts of activities beyond their own

Guy Warner and Jack Woods' trilogy of books about Northern Irish airports were among the first of many published through the Society over the years.

experience. A member could become a restorer, a researcher or a magazine editor. Members have served as librarians, building renovators, event planners, public speakers, cleaners, photographers and tour guides. Some are licensed pilots, plastic model builders, medal collectors or amateur radio enthusiasts. Some brought to the Society their own specialised backgrounds, but most learned through doing. Others have come to the group with little but enthusiasm, and many have revelled only

Guy Warner (here signing a book for member John Barnett) is the most prolific of a number of authors the Society has nurtured over the years.

skill. Each effort, valuable on its own merits, also reflected the constitutional objective of public education, another means of reaching out beyond the Society's internal interests. Many members of the public, touched by that outreach, signed up to enjoy the organised trips and special flights which remained staples of Society activity throughout the late 1990s and into the new millennium. At once educational and entertaining, the excursions provided members with an opportunities to socialise, spot or photograph aircraft and to just enjoy a relaxing day or two away from the daily grind of work and family duties.

in the arrival of their *Airmail* each month. Some have never even visited the collection, but all have been welcomed.

The first opportunity afforded to the members of the Society (beyond plane spotting, at which most were already dab

The intrepid visitors were astounded to witness five Antonov An-124s.

hands) was writing. *Ulster Airmail* was in constant need of movement logs and articles, and members were invited to contribute anything they could. Here was an outlet for enthusiasts who had little or no writing experience to have a go and develop a new

On June 14th, 1998, for example, members boarded yet another Dublin to Shannon flight, this time on a McDonnell Douglas MD-11 (N272WA). Fully expecting to see the usual uninspiring gathering of airliners on Shannon's ramp,

Members couldn't believe their luck when a routine trip to Shannon turned into a guided tour of the colossal Antonov AN-124 freighter, at that time second only to its sibling the AN-225 in cargo capacity.

For many the dawn of budget flying opened up a world of destinations, but for the Society it spelled the end of discounted Society trips.

the intrepid visitors were astounded to witness no fewer than five Antonov An-124 freighters gracing the tarmac. Each of these behemoths was capable of lifting a 250-ton load, so the sight of so many was incredible. Some quiet discussion between the airport police and the Russian crews yielded an amazing result for the wide-eyed visitors: guided tours inside the bellies of the beasts. The members were brought down to earth by the approach by two of the crew, who asked if they had been photographing the transports. "We thought that maybe this time we had overstepped the mark," Garry Adams wrote. Crestfallen, the photographers bowed their heads and confessed.

Then came the kicker: "They told us it was best not to do so here because the light was better on the other side of the apron!"

Before they returned north, the

members were in for one last treat for their benefit alone: a full demonstration by the crew of the Antonov's 20-wheel unique undercarriage, which can "kneel" to provide easier access of cargo into its hold. Those who were there still talk about that excursion all these years later.

A few of them may still talk as well about another flight only three months afterward— one that would spell disaster for the Society's organised, cheap flight experiences. On September 18th, an EasyJet 737 landed at Aldergrove on its inaugural Luton-Belfast-Luton route. At only £28 return, the service heralded a new era of budget travel from Northern Ireland's airports.

The trips on offer for members became notably fewer as the 21st century approached, the first sign perhaps of the impact of budget airlines. The steady public acceptance in the

Steadfast Society supporters Aer Lingus offered to bring their historic DH-84 Dragon, *Iolar*, to Langford free of charge.

coming years meant the decline of tasty deals offered through Society trip organisers.

"We could not compete, pricewise," Ray Burrows explained some years later. "Why should anyone pay £100 for a Society trip when they could fly themselves there and back for a fraction of the cost? In some ways, it was sad; we never really recovered the good old magic and craic of those trips."

But one special offer to the Society ignored low-cost commercial travel, and opened a delightful window to the past. Steadfast Society supporters Aer Lingus offered to bring their historic DH-84 Dragon, *Iolar*, to Langford free of charge. Not only were flights in such vintage biplanes highly coveted, but this one had a unique connection to the island, adorned as it was in the livery of Aer Lingus' first aeroplane. (The original *Iolar*

As trips abroad were curtailed, members could still travel locally. Several trips were made to the Irish Air Corps home at Baldonnel, growing a relationship that would pay dividends in the years to come.

was lost during the Second World War.)

True to their word, the Dragon touched down at Langford Lodge on September 2nd, 2000. And the day provided another surprise: the arrival of a Yak 18 in a beautiful "tiger" paint scheme. The pilot was Warwick Creighton, who arrived with his wife, Pat. And she just happened to be the daughter of the first pilot to fly the original *Iolar* in 1936! The Creightons offered rides to any member who wanted to have a go in the Yak.

When it came to Society trips, the annual pilgrimage to the Leuchars Airshow was not to be missed.

It was a terrific day's flying for the Ulster Aviation Society, said Ray Burrows: "A lot of people will remember that day for their flight in the Yak, screaming round Lough Neagh at a hundred feet, frightening the life out of a bunch of canoeists!"

When it came to Society trips, the annual pilgrimage to the Leuchars Airshow was not to be missed. The year 2000 brought fantastic weather, and with it the familiar tone in *Ulster Airmail* of Ray's light-hearted recollections of the day itself: "Nothing but blue skies, God bless the met man….It took some time to get the noise of jets out of one's ears….I made the mistake of visiting the front of the coach where two of our more mature members would not let me depart without my fair share of something called 'Cream of the Barley.'"

They were building memories of another enjoyable trip by all to Scotland's east coast.

The format of the excursion evolved in tandem with the rise of the budget airlines. Leuchars '02 would see the

Aer Lingus keep their history alive through Iolar, *and were only too happy to bring it north to visit the likeminded members of the Ulster Aviation Society. Here it is in 2012 when it visited the Society for a second time.*

participants booking their own transport for the first time since the annual pilgrimage started in the 1970s. The cheap seats presented a better option than the usual group ferry booking could.

Still, there was entertainment to be had at home. A fundamental fixture of Society membership was the monthly meetings, which continued with a wide variety of speakers visiting CIYMS, speaking on such subjects as hot air ballooning, the Jet Heritage Museum and air-to-air refuelling.

The year 2001 got off to a wonderful start as Paddy Crowther entertained the crowd with exploits from his days as a pilot with the Royal Air Force, Queen's University Air Squadron and later with Shorts. The latter part of his career provided the highlight

of the evening, as he recalled such problems as a basic one caused by the large, flexible fuel containers installed for ferry flights. That might have prompted some bladder tank humour.

"(They) blocked the way ... to the only toilet at the rear of the aircraft!" laughed Paddy.
The *Airmail* meeting report happily labelled Paddy's contribution "a cracker of a night."

Much of the success of many such talks could be attributed to their first-person character. The guest speakers were often, like Paddy Crowther, at the front line of experience in aircraft and events which loomed large in aviation history. That was the case, for example, in June of 2003 when ex-Concorde captain John Hutchinson addressed over 60 members at Langford

Shorts test pilot (and later Society trustee) Paddy Crowther's storytelling aptitude provided one of the most memorable talks in Society history.

John Hutchinson had the room in the palm of his hand as he delivered his detailed and honest appraisal of the 2000 Concorde crash.

correspondent. There was no hiding his sadness at the impending retirement of the high-speed icon, mere months away from its final flight. Even Guy Warner, a member with cool, detached demeanour as a respected historian, admitted in the *Ulster Airmail* to "a certain moistness about the eyes." Others agreed at the time that John's lecture was the best the Society had ever hosted.

The *Airmail* itself, through all the Society's ups and downs, remained an unfaltering, common connection among members near and far. Gary Adams had taken over as editor in January of 2001, allowing Ron Bishop to take a well-earned rest. Gary immediately began to add his own distinctive touches to the magazine, including a notice board of smaller Society news items—recognition that his readers wanted to know

Guy Warner admitted to "a certain moistness about the eyes."

on a regular basis what their organisation was up to. And starting that June, the entire publication had undergone a facelift, with colour covers and more photographs inside.

"We can now add a visual dimension to both the articles and to reports of aviation events and movements," he humbly reported.

Gary's term as editor was brief—only two years—but it marked the biggest step forward in the magazine's history since the A5 format introduced by Stephen Boyd.

His work had been aided by a global phenomenon which had gradually snuck up on the organisation: computers and the information technology (IT) programmes which ran them and served their users. It had taken the leadership, as it did society at large, years to recognise and adapt to the significance of IT and its ever-changing potential to benefit the membership.

Michael McBurney had the honour of bringing the Society's website into the 21st century. The address was changed in July 2001 to www.ulsteraviationsociety.co.uk and the site adorned with pictures submitted by members of past and present Society activities Thoroughly researched articles on Langford Lodge's remarkable history and other items of interest to aviation enthusiasts were added in the hope that the site would become the go-to destination for visitors seeking information

on aviation in Ulster.

As for travel destinations, by 2003 trips abroad had become increasingly difficult to organise in a collective fashion, though the indomitable Guy Warner did manage to arrange one experience flight for that year. That venture took 13 members aboard a Jet Magic Embraer ERJ-145 on August 13th to Cork City Airport— unexplored territory for the Ulster Aviation Society.

To fill the gap of fewer flights, the organisation had sought opportunities closer to home. To that end, Ray Burrows had begun guided tours of Belfast International Airport's air traffic control centre on March 19th. Eight months later, members enjoyed a rare guided tour of the Queen's Island facility of Bombardier Belfast (Shorts), followed by a talk from site manager Michael Ryan.

Tours, trips, books, meetings and magazines may have made up the backbone of the Society but its growing heritage collection was rapidly becoming the group's 'shop front.' It was firmly established at Langford Lodge and its popularity was growing, so the Ulster Aviation Society management committee began to organise a string of annual open days for the public. They took several forms over the late 1990s and into the early 2000s but the group always aimed high, each event more ambitious than the last. They were evidence of the Society's focus shifting wider, to include

The Society's reputation was growing and it showed at the annual heli-meets. The support of the local armed forces was clear when in 2001 they sent four helicopters to the open day.

the general public as well as the group's membership.

Things had started small. August, 1998 had seen the arrival of ten helicopters at Langford for the inaugural Ulster Aviation Society "heli-meet." First on the ground was Lynx XZ662 of the Army Air Corps' 655 Squadron—the steadfast supporters who had played a major role years before in the Society's aircraft retrieval activities. For the members, it was a day to inspect the aircraft and meet the crews in person. For those crew members, it

was an opportunity to add a rare airfield to their logbooks. UAS volunteers laid on a barbecue and organised a "Bushmills ballot"—a chance to win two bottles of whiskey from the latest sponsor, Old Bushmills Distillery.

The following year was a highlight for members and the public alike. The British Aviation Preservation Council selected nine days that spring as Heritage Week. The aim was to raise awareness of aviation history and nationwide endeavours

On a dead-end road, the Society could never rely on passing trade, though they made every effort to announce their presence.

As the collection grew, Langford became the Society's 'shop front,' though its quiet location on a dead-end road and proximity to Langford Lodge Engineering meant only pre-arranged visitors could cross the threshold.

By the late 1990s visitors to Langford had a series of interpretive displays to peruse alongside the aircraft that made up the bulk of the collection.

Lodge. He had the benefit of two points of view—that of the pilot and that of his later career as the BBC's aviation to safeguard it. For its part, the Ulster Aviation Society chose to showcase its place as the home-grown curators of Northern Ireland's aircraft heritage. The celebrations culminated in the second annual Langford Lodge heli-meet on May 2nd.

The event included a display of 350 vintage cars and a special, one-day radio station operating from the Langford tower, hosted by Richard Ferris. Timely advertising, coupled with free entry, resulted in no fewer than 1,500 visitors on the ground with nine aircraft coming and going over the course of the day. Besides escorting hangar visits, the heritage members could show off their exhibits in the parachute building, where they had toiled in recent years

carefully arranging aviation displays. The Society hosted a third successful heli-meet in 2000 and, though it didn't yet have a rotorcraft in its possession, the heritage team were dropping hints that the situation might soon change.

The success of such events was evidence to the Society's leaders of potential for further public outreach. Aside from carefully-planned open days, visits to the collection had become a regular occurrence, often involving school classes or special groups like the Deaf Christian Fellowship. It was a pleasure to meet interested visitors and the group prided itself on fulfilling its charitable remit of educating the Northern Irish public in the story of flight.

Such visits were cut short early in 2001 by the foot and mouth crisis. Government officials considered that cattle on

The UAS made a specific effort to celebrate aviation across the island of Ireland, and to that end invited crews from the Irish Air Corps to join their annual heli-meets.

nearby pasture land might be a threat to health. When the site was finally reopened in May, the committee was eager to put the Society back on the public radar. The European Heritage Weekend was the perfect opportunity to do so. Billed as a 'Vintage Extravaganza,' the site opened its gates on September 8th to over 1,000 visitors who wandered among vintage cars, motorcycles and buses, chatted to the aircrew of all four military rotorcraft operating in Northern Ireland (Wessex, Puma, Gazelle and Lynx).

Not content to make each year's event a repeat of the last, the 2002 committee busied itself with preparations for its biggest open day yet. This time they would stretch beyond a helicopter fly-in and plan a full-blown open day at Langford Lodge. The theme would be Wings and Wheels, paying tribute to the thousands of American soldiers and civilians who had been stationed at Langford during the Second World War. The day was set for

June 16th. This was a strategic choice given that the Ulster Air Show was to take place on the preceding day.

"Co-operation with the organisers has opened the doors for a great event to be held at Langford Lodge," explained Gary Adams, "to commemorate the 60th anniversary of the arrival of the Americans." A corresponding series of *Airmail* articles researched by USAAF enthusiast Ernie Cromie provided members with some historical context for the celebrations. The joint weekend occasion would be the greatest single effort of the committee to date, he noted: "Months of planning, more phone calls than I wish to remember and more meetings that we have ever held before. Everything is falling into place."

But then…collapse! Three days before the air show, chief organiser Jeff Salter announced its cancellation, due to a forecast of torrential weather and the likelihood of flooding.

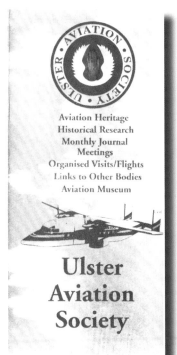

Professionally printed leaflets showed that the UAS was serious about their quest to open a fully-fledged museum.

Still, he promised to help the Society's Wings and Wheels event as best he could. As if to reinforce his support, Jeff himself was the first to touch down at Langford in his Piper Cub, followed closely by a show highlight, Grumman Albatross N7025N. The flypast was limited, of course, but was aided by agreement from 230 Squadron, a Puma helicopter unit, to route all its air traffic via Langford. The public was soon thrilled as the Chinook, CASA 235, Puma and Grob glider wheeled overhead in turn. On the ground an impressive turnout of military and classic vehicles delighted the crowd. Not to be outdone, the Ulster Aviation Society presented the star of the show: Buccaneer XV361. It was rolled out, jacked up and connected to ground power. It performed a perfectly choreographed routine of wing folding, undercarriage lifting, air brake deployment and bomb bay rotation.

Jeff Salter touches down at Langford Lodge, opening the 2002 open day. He strongly supported the event despite the Ulster Airshow being rained off the previous day.

It was thanks to Jeff Salter that visitors to the 2002 open day had the opportunity to see Grumman Albatross N7025N in the flesh.

The Lurgan Concert Band provided a soundtrack to the day, covering hit music of the legendary Glen Miller, who had led his band in a concert at Langford Lodge in August of 1944.

On June 17th, UAS Chairman Ernie Cromie addressed the congregation at Langford's picturesque Gartree Parish Church at a Sunday service of thanksgiving on the occasion of the site's 60th anniversary. *Airmail* described the site beautifully: "Gartree Parish Church is now much as it was sixty years ago, an oasis of peace and worship surrounded completely by the runways, aprons and hangars of the now silent airfield at Langford Lodge."

Fundraising was an ongoing challenge. Colourful Christmas cards like this one featuring G-BTUC went some way to filling the coffers.

The former base offered a quiet location as well for a television documentary, shot in November, 2002. It featured Sir Philip Foreman, retired chairman of Shorts Aircraft, and his part in the development of the company's seminal Skyvan. He was right

at home on the flight deck of SD3-30 G-GDBS, a fellow retiree.

Sir Philip's visit and his continued encouragement gave the Society's committee pause for thought. Here was an ideal figure to represent the organisation, a potential emissary to Northern Ireland business and the community at large. Knowledgeable and respected but down-to-earth and approachable, his donations of airframes in the past were clear evidence of his support. He would make an ideal patron, said Ray Burrows, who himself had a Shorts background as a young engineer.

"(Sir Philip) got the reputation of being a very outspoken man, very forthright," he said, recalling the boss's visits to the company's shop floor when Sir Philip was managing director. "Whenever he came, you just turned and ran and hid in the corner," he laughed. "However, as patron of the Society, he could not have been nicer.... He was a complete gentleman."

Agreement from the membership at the 2003 AGM sealed the deal and Sir Philip Foreman accepted the honorary position that August, one that he retained until his death in 2013.

"He was a very good ambassador," said Ray. "He would always mention the Society when he went out to meetings…He would ask people, 'Do you know about the Ulster Aviation Society?' And apart from that, he was

very generous. Even when he died, his widow, Lady Margaret continued to send a cheque at Christmas time." It's a credit to the Society committee, given the monumental effort involved in organising their yearly open days and fly-ins, that they had continually set their sights higher. Each year's event had to be bigger and better than the last, and 2003 was to be no exception. That year's open day, marking the centenary of powered flight, was to be a two-day event for the first time in the Society's history. The organisers set September 20-21 as the weekend, coinciding with European Heritage weekend, and helped along with a grant of £1,900 from Antrim Borough Council. The occasion was a hit, with almost 3,000 visitors passing through Langford's gates. As before, the ground was graced with vintage vehicles while at the hangar the Society Buccaneer again flexed its wings. Indoors, pilot Peter Lock—our Wildcat's retired pilot—joined another RAF 502 Squadron veteran, Dickie Spencer, to regale guests with yarns about their flying exploits, to music provided once again by the Lurgan Concert Band. This time, it was the turn of the visiting aircraft to out-do the Society effort as Spitfire PT462 touched down on the airfield. The usual complement of local helicopters completed the visiting lineup and good weather topped off another unforgettable weekend at Langford Lodge.

A unique feature of Langford Lodge was Gartree Parish Church. Nestled among the hangars it was a picturesque and peaceful oasis on a site more used to the howl of aircraft engines.

Former chairman of Shorts, Knight of the British Empire, respected engineer and loyal and generous supporter of the UAS: Sir Philip Foreman was an ideal patron for the group.

Such events also had proven by this time to be ideal recruiting occasions for new volunteers. John Martin was one such helper, and has recalled the open days in the early 2000s as a neophyte car-parker with affection.

"It's engaging the public," he said. "The first impression that people get is when they come through the gate and meet the people who are doing the car parking and whoever is taking their admission money and whatever else. If that's not done right, then you're on a downer."

The experience, plus his aviation interests since childhood, hooked John into membership that has taken him to a valued place on the Society's management committee. It was a progression that included regular hangar work once he'd become acclimatised. "It got into my blood and I started going over on Saturdays and that became a regular feature of my life."

It didn't hurt, either, that his career in IT was of direct benefit to the Society's website improvement.

Along the way, he was learning, aided by volunteers like Leonard Craig and Neville Greenlee—the latter a winner of three consecutive BT Community Champion awards by this time. Neville donated the cash award to the Society, a gesture typical of many members throughout the years.

Both men were frequently

The Society raised funds anyway it could. Chairman Cromie led an effort to collect and crush metal cans which could be exchanged for cash.

in action at the Langford property, restoring exhibits but— becoming more important—also assisting in repairs to the hangars themselves. The group may have had a roof over their heads, but it left much to be desired.

"Every time there was a storm, part of the roof disappeared and we never found it," lamented Ray Burrows. The team resorted to suspending plastic sheeting from the trusswork of the failing ceiling to try to weatherproof the wanting hangar while shielding exhibits from dust and debris. New windows had been fitted to the western wall and the painting of the floor was almost complete. The old parachute shed on Langford Lodge was also made available to the Society and members wasted no time in putting it to use as an area for displays, lectures and modest kitchen facilities. The old control tower, though of secondary interest to the

Society at this point, was spruced up with new paint inside and graced with no less than 400 new window panes. "The amount of work we put into Langford Lodge, I couldn't put a price on it," Ray said years later—his voice still, after all that time, betraying exhaustion from the team's efforts.

While others busied themselves with the upkeep of the hangars at Langford Lodge, Ernie Cromie was a driving force behind the exhibits. The Society records from that time note the steady progress: "The chairman still secretly adds photographs to the various displays, which means you have to look at them each week to see what's new....The new boiler has been connected up and working..... An absolutely spotless library facility has to be seen to be believed...."

The latter project was home to over 500 books in various categories including biographies, novels, histories

Leonard Craig (in hi-vis) looks on as Neville Greenlee (in cockpit) readies the Buccaneer for its routine. Leonard and Neville are typical of the unassuming volunteers without whom the Society could not function.

Crowds look on in amazement as Irish Air Corps Dauphin 245 touches down at Langford. The Irish Air Corps are strong supporters of the Society to this day.

Peter Lock formed a close bond with the custodians of 'his' Wildcat and returned to Northern Ireland many times over the years. In 2003 he joined the open day team, talking to visitors about his wartime experiences.

As the Society increased its presence at Langford its tenancy was extended to include the old control tower. The building was completely re-glazed.

The vintage Vampire looks sleek and modern in comparison to its surroundings. During their tenure at Langford the volunteers put as much effort into the upkeep of the site as they did the restoration of the aircraft.

Booking a B-17 for the 1995 open day was a coup for the UAS, only to be repeated in 2003 when the unmistakable roar of a Rolls Royce Merlin engine echoed around the site, signalling the arrival of Spitfire PT462

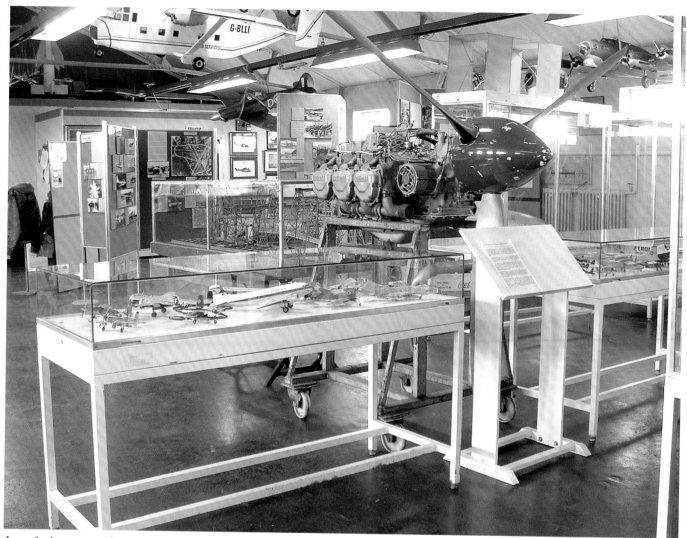

Langford was more than just home to a collection of aircraft, it was a museum, and volunteers like Chairman Ernie Cromie used it as a training ground to develop skills in exhibition design.

and reference books on a wide variety of aircraft types, operations, radar, wireless communications, space exploration etc. One of the visitors to Langford Lodge at about that time was a Second World War veteran, Fred Jennings. He likely had no idea how significant he would become to the Society's future connections with radar and communications equipment and, above all, books.

Building renovations and attractive displays, as important as they were, comprised only part of the group's Langford activities at that point. The leaders were also busy acquiring aircraft. An odd little amphibious biplane called a LeGare Sea Hawker (EI-BUO) was donated with only 20 hours of flying time and has rested for years, unrestored, in the UAS hangars in a what-do-

One of the visitors to Langford Lodge at about that time was veteran Fred Jennings.

we-do-with-it-now state.

Pilot Billy Reed donated a Chargus Cycle hang glider at about the same time as the Society acquired a Rogallo glider. None of these may have struck heritage members with great bursts of enthusiasm. However, a gift is always appreciated and the types helped to provide a balance against the mainly military aircraft in the inventory up to then. Retired *Flypast* editor Ken Ellis had warned the group in his address five years earlier that an overlooked aspect of flight in modern museums was general aviation.

Those civil aircraft acquisitions addressed that problem. It helped as well that Shorts had donated several large and rather impressive models of their commercial designs. The company (purchased by Bombardier in 1989) enhanced its reputation for generosity in 2001 by selecting the Society as the ultimate home for Tucano G-BTUC, the Embraer-built development aircraft which had preceded Shorts' production models. Coincidentally, it had been one of UAS member Paddy Crowther's charges when he had served as a Shorts pilot.

As the demands on the volunteers increased the heritage subcommittee had been inaugurated with a mandate to manage the specifics of what was now called the heritage centre at Langford Lodge. Though still responsible to the Society's management committee, it had a healthy measure of independence. It would supervise volunteers, look after electricity bills, health and safety, visiting groups and restoration projects. This would lighten the load on the parent committee and also give the heritage volunteers a sense of ownership of their branch in the larger organisation.

By 2003 the Society had finally got its hands on a

The UAS strong relationship with Shorts, with Sir Philip Foreman as an intermediary, came good again in 2001 when the company selected the UAS as the ultimate custodian of prototype Tucano G-BTUC.

Civil and military, homemade and factory built. But with Harold Hassard's loan of his own R-22 the Society filled the most obvious gap in its collection: rotorcraft.

The donation of several hang gliders and home built aircraft like the unusual Goldwing (shown here at Long Kesh) added another facet to the UAS collection: light aviation.

rotorcraft for the heritage collection. Though not the Wessex that the committee had been tirelessly chasing, the bright red Robinson R-22 (G-RENT), loaned by owner Harold Hassard, was in fine condition inside and out. Moreover, its arrival provided more variety for interested visitors. Military, commercial and general aviation were all represented, and now a rotary-winged type could be added to an ever-diversifying collection. An early Christmas present from Castlereagh Borough Council in the shape of an RAF Jet Provost T3 trainer (XM414) made for an upbeat

end to 2003. The airframe was offered to the Society on long-term loan after plans for a flight experience centre project collapsed.

Day by day, the collection at Langford was starting to look more and more like a museum. It had grown not only through donations of aircraft, but a steady stream of associated items—uniforms, photographs, scale models, instruments, tools, part of a propeller and other bits returned from the wreckology treks of years before. Much of that generosity was due to hangar visits and occasional

The addition of Jet Provost XM414 helped to build a relationship with Castlereagh Council, and added another iconic type to a rapidly growing collection.

publicity, with people seeking to know whether the Society was open to receive such donations. Ernie Cromie recalled one especially valuable contribution which involved the group's Wildcat.

"That was the remote compass, which would have been located in the fuselage of the aircraft. The individual had been at the site and recovered it, and…it would have been in the 1990s that it was given to us," he said.

Financial donations—many of them through hangar visits—were always welcome and were enough to sustain heritage operations and the costs of open days. A sizeable gift of £230 from visiting RAF 230 Squadron aircrew, for example, was provided on the strength of the collection being a

Not just limited to aircraft, Langford was home to uniforms, logbooks, photographs, engines and much more besides. Here Ernie Cromie shows a visitor one of many aircraft models in the UAS collection.

Any Wessex would do, but XR517 was special. It had served in Northern Ireland with 72 Squadron - the perfect example for the UAS collection.

museum in progress. Still, big purchases required big money and the incoming donations just weren't going to cut it. At the turn of the millennium, Ray Burrows had made an appeal to the *Airmail* readers to consider contributing to the tellingly-named Wessex Fund. By the end of 2001, that fund had reached £5,000—all of it in donations. The outstanding generosity of the members, both financially and in the time they gave up for the Society was noted in the *Airmail*: "Langford ... is a flagship for voluntary charitable groups, such as we. How else could we have saved such a sum whilst still improving our all-around facilities?"

The UAS was no stranger to fundraising, and with its sights set on a Wessex it was time once again to drum up financial support.

Initial enquiries into the purchase of a surplus Westland Wessex helicopter had been met with cold responses from the RAF, but the heritage members were not disheartened. The Society took the offer of a free Wessex simulator and sold it for a meagre profit. In January of 2002 the RAF announced its intention to retire the type. After 30-plus years of service in Northern Ireland, it was slated to disappear in the coming March. The stage looked set for another landmark acquisition for the Ulster Aviation Society.

Retired or not, the coveted helicopter was not forthcoming and 2003 passed with barely a whisper of a Wessex. Inquiries to the Ministry of Defence were frustrating, but ended with a sealed bid from the Society for £12,000 for a veteran machine. It wasn't high enough and the MoD sold it elsewhere for £18,000. However, an offer for a private sale in Shoreham, West Sussex, appeared enticing. Negotiations with owner Dick Everett concluded with a Society bid of £8,100 plus VAT—less than half the MoD price. But what shape was the aircraft in? Ray Burrows contacted a Society friend, Mark Fitzherbert, in England, who offered to check it out. His inspection, said

Ray, proved positive: "You've got to buy it! It's absolutely complete except for the engines; you'd think it had just landed. You've got to go for it!" And the heritage team went for it, with full committee backing.

The *Airmail* boasted: "This particular Wessex is XR517, which served with 72 Squadron at Aldergrove for 21 years and is in very good, almost complete condition." There was much to boast about. Besides its years of yeoman service to the Army during the Troubles, XR517 had performed occasional civil duties when called upon. One outstanding example was its heroic performance with other 72 Sqn aircraft in the rescue of 128 passengers and crew from the *Antrim Princess* on December 9th, 1983. The Sealink ferry had caught fire during high winds and seas off Larne. She had lost power and was heading for a nearby rocky shore when the aircrews performed their hazardous operation. Luckily, the ferry managed to anchor just off the rocks and survive the ordeal overnight ; she was towed into Belfast for repairs the next day.

Twenty-one years later, with the Ulster Aviation Society knocking at its door, another ferry company returned the favour. P & O, with a nudge from Maxwell Freight, provided a discounted fare to bring the helicopter back to Northern Ireland. Days of frantic phone calls followed, to ensure adequate road transport and unloading facilities.

Finally, on March 30th, 2004, Wessex XR517 entered the Langford Lodge compound. Ray made special note at the time of the assistance received by the Society's neighbours down the road: "We owe Terry and Tommy Maxwell a huge vote of thanks for going ahead and arranging the ferry crossing through their own company." Special thanks were owed to seller Dick Everett, who had supervised the loading at Shoreham, including removal of the undercarriage and two tail rotor blades to enable the helicopter's transport across the Irish Sea. In the end, back again in Northern Ireland, the aircraft required little in the way of restoration.

Celebrations were short lived however, as the homecoming of XR517 coincided with a much more unwelcome arrival. This one came by post: An eviction notice in March to leave Langford Lodge.

In hindsight, it may seem odd that in the midst of this calamity the committee found time to arrange another stellar event. But this tenacity was typical of the Ulster Aviation Society.

Langford Lodge management was now discouraging public events, but the Society secured an alternative, back at its Aldergrove roots. A huge marquee, with welcome assistance from Antrim Borough Council, went up at the airport, and 1,350 visitors flocked to attend a special event on August 14th. It was

Finally, after years of fundraising and rejected bids, a Wessex passed through Langford's gates. Ray Burrows couldn't hide his excitement.

Finally, on March 30th, 2004, Wessex XR517 entered the Langford Lodge compound.

a nostalgic spectacle of military vehicles and fly-pasts by two Piper L-4 liaison aircraft in World War II U.S. military markings. The highlight was a big band concert by the John Miller orchestra, fronted by the nephew of the great Glenn Miller—he of Langford Lodge reputation. In fact, the audience included at least three women who had danced with Glenn Miller himself in 1944, and a veteran who flew in from Michigan just for the occasion. It was a terrific event for an organisation in crisis.

BT29 4RI
TELEPHONE 028 9445 1400
FAX 028 9445 2161

MARBUR PROPERTIES (NI) LTD

APM/JK/L01270001

26th March 2004

Mr Ernie Cromie
Chairman
Ulster Aviation Society
27 Woodview Crescent
Lisburn
BT28 1LE

Re :   **Hangar 6**

Dear Mr Cromie

I regret to inform you that the Company has now decided to terminate its present arrangements with the Society, whereby the Society has a licence to use four bays in this hangar, together with the parachute shed and control tower, (respectively Buildings 6, 10 and 30 as shown on the attached Site Map) now that the hangar has been listed as a building of special architectural or historic interest.

Accordingly, this letter constitutes notice of the termination of the Agreement by which the Society has a right to use the hangar space, parachute shed and control tower, with effect from 31st October 2004. I trust that this will give the Society sufficient time to find a new home for its collection of aircraft. Please also note that the Company will not grant consent for Langford Lodge Aerodrome being used for any special event. during the period of notice, and has reached their decision, for safety reasons, but with regret.

Yours sincerely

Director

Possibly the most infamous document in UAS history: this was the letter that ended the Langford dream and entered the Society into its most tumultuous period.

But a crisis it was. Attempts to save the collection's presence at Langford Lodge had hit a wall with the landlord's property management. The committee's attention now turned to the inevitable hunt for a new home. Meanwhile the effort to dismantle the Society's substantial selection of aircraft and artefacts began as pleas were made for help: "Every member is more than just welcome. You are needed." Those with practical know-how took on the disheartening task of reducing their aircraft collection, restored over the preceding decades, to pieces while the management-minded continued the unenviable search for a new site. A wish list was laid out. Space was of course the main concern, not least to house the collection as it was but with the potential to expand in the future. A site with an accessible runway would be ideal, with access to major roads being an acceptable alternative. Costs and leasing arrangements were factored in. Desired though it was, it was doubtful that a home with some heritage like Langford Lodge would be forthcoming. The year 2005 started in silence—a stark contrast to the optimistic swagger of just a few months earlier within a Society that could do no wrong. There was still no news for the Society's rank and file about the uncertain future. In fact, many appeared to be dispirited, as membership numbers fell below 300 for the first time in years.

"Every member is more than just welcome. You are needed."

There was little the leadership could do to assuage the concern, and it showed in Ray Burrows' trite summary to the members in January: "There is no doubt it would help if we knew where we were going."

His frustrated observation was accompanied by a pitiful

A change from the annual open day, in 2004 the Society invited John Miller to Northern Ireland to recreate the sights and sounds of wartime Ulster.

80-year-old Warren Bradley travelled all the way from Michigan for the John Miller concert, and was given the warmest of welcomes.

plea for members to bring cardboard boxes and bubble wrap to Langford Lodge.

So there it was, a pair of major, simultaneous challenges, if the heritage collection—and likely the Society itself—was to survive: Pack up everything in the coming months and at the same time find a suitable place to put it.

While helpful hands began to dismantle aircraft, other hands dug deep and donated cash to fund the move, and the committee sent scouts to find a new home. They went to Carrick Council, which suggested a field at Bentra, near Islandmagee. It was large enough but would need a big, new building; that meant more money and, besides…Bentra? Too far from anywhere. Perhaps the old Courtaulds rayon fibre factory at the edge of Carrickfergus? A site survey showed a good location and a choice of big buildings of suitable size. But costly renovation would be needed, with likely a high rental. Not a chance. Besides, the experience with the Newtownards council years before had left a bad taste, said Ray.

"I guess most of us, including Ernie, thought, 'Well…what chance is there of getting another council to help us?' And we decided we'll just plod on ourselves and try to do things as best we can." So they did. A strong letter went to Aldergrove International Airport, the Society's birthplace, but the offer only of a green field site

had serious cost implications. At about that time, in the spring of 2005, a member's sister who worked in the Northern Ireland government overheard planners discussing the derelict Maze Prison at Lisburn and a pair of adjacent wartime hangars. Political palaver had brought a halt to the demolition of the prison while the site planning was going on, but the hangars themselves were just a question mark. Ray recalled the chain of events.

"She says, 'I know exactly what you could do with them, 'cos my brother works with an organisation with aircraft, and they've been told to leave their premises.'"

That little story has grown in the telling, with minor variations, but the point was clear: There was a group with a ready-to-go museum, and what was a hangar or two without aircraft? Contact was made with Tony Whitehead, advisor to the government's Strategic Investment Board (SIB), and the response seemed favourable. His team met with Ernie Cromie, Ray Burrows and Paul McMaster and gave them a tour of the premises, which crouched mere metres away from the grim, hard face of the empty prison. The size and location looked right—ideal, in fact, considering the site's history. It had been an active RAF station in the Second World War. The massive complex with its sad and bitter past would hardly make an ideal neighbour, but Tony Whitehead indicated a rental deal on the

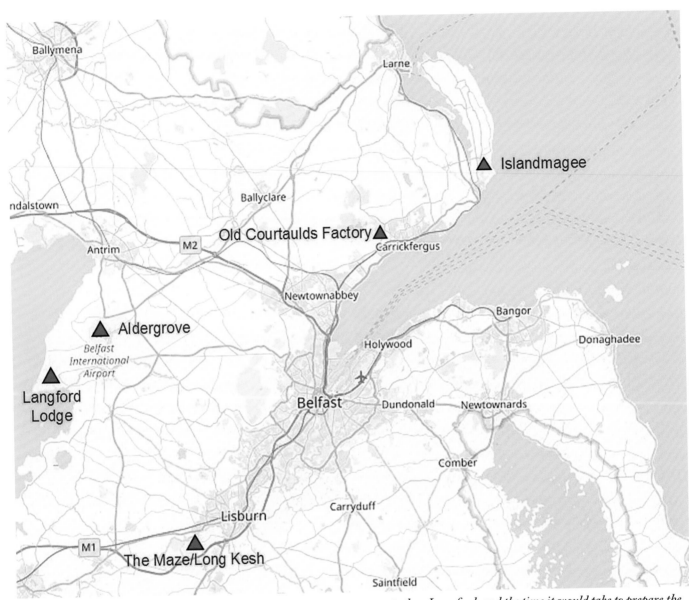

Islandmagee was floated as one potential new site, but it was even more remote than Langford, and the time it would take to prepare the site ruled it out entirely. As time went on, the Maze/Long Kesh emerged as the only viable candidate.

hangars could prove to be very reasonable—at least, financially.

From the air, the remnants of the infamous prison disguised what it once was—RAF Long Kesh, an ideal heritage foundation for an aviation museum. Still apparent in places, its perimeter road follows the airfield's

What was a hangar or two without aircraft?

peri-track, the tail end of a runway and the corrugated metal hangars themselves, still standing as if in bold defiance of bulldozers and breezeblocks.

From the government's point of view at the time, an aviation museum could feed right into an ambitious master plan which

had been commissioned for the entire, 360-acre Maze/Long Kesh Development Site. Work on that plan had been initiated in 2003; it was due to be released in January, 2006. The dreamers behind the planning saw in the site a great opportunity: empty land, a certain amount of infrastructure, proximity

Built in 1940/41 for the manufacture of components for Short Stirling bombers, this pair of T2 hangars are the only significant structures remaining from the aerodrome days.

to Belfast—a location ripe for development. For many, it represented an opportunity to leave the troubled past where it belonged and make the most of the province's optimism of the day. For others, however, the notorious property remained an uncomfortable, even painful reminder of terrible times past. So, what to do?

The Society and the SIB began negotiations, but very discreetly, away from the curious eyes of the membership. The whole site was politically sensitive; loose talk at the wrong time could scupper the discussions and perhaps the whole project.

Anyway, the volunteers were occupied with other duties, namely preparations to leave Langford Lodge. There was a curious sense of opportunity about the pending departure from Langford. While the aircraft were being dismantled, the work created a chance to spray corrosion protection on many of the internal parts.

The membership was rallying to save their organisation. It was not the first time such a collective effort was made, but it was certainly the largest and the most urgent.

Not that the Society had frozen all its other concerns; for example, Michael Bradshaw (acting editor of the *Ulster Airmail*), was stepping down. Member Graham Mehaffy could see that Society Chairman Ernie Cromie was under stress, as was evident in an *Airmail* address to his flock: "The last fourteen months or thereabouts have amounted to the most vexatious period in the history of the Society." Graham volunteered for the editor position in the midst of that vexation; his decision was no doubt an overdue tonic for the organisation.

"They were stuck, really stuck—in dire need," Graham said, recalling his qualifications and noting he had done considerable graphics work for

Leaving Langford

The eviction notice to the Society was not entirely out of the blue. A six-month rolling lease of the Langford Lodge buildings had not been ideal from the start.
Said Stephen Boyd: "I always felt very uneasy that we would be told, 'Sorry, you've got to go.'"

Marbur Properties, site managers, said the company was concerned about public access to the site and the increasing number of visitors. The company's representative reminded UAS leaders that originally there would be only two or three aircraft, but additional aircraft and displays had heightened the collection's popularity. As well, the doubtful condition of the hangar suggested a potential health and safety risk to visitors, and thus to Marbur itself, said an internal Society report at the time.

Paul McMaster worked harder as Society secretary than he ever imagined he might. Letters flew back and forth, meetings were held, alternatives debated. Changes to the lease and/or its limited tenure were out of the question, said Marbur. Greater permanency for the collection would restrict the company's options, it claimed. Pleas that the best collection of heritage aircraft in Ireland might be lost fell on deaf ears. Tourism potential was disregarded. Even a Society offer to pay for upkeep and renovation of the leased buildings was not enough.

Public access would be refused after October, 2004. Marbur's only concession was a one-year extension of the eviction notice to October, 2005. Once again, the Ulster Aviation Society would have to find a new home.

"Yeah, I was a bit peed off about how we'd been told to leave, not given a chance," said Ray Burrows, his voice years later tinged with sadness. "Langford Lodge was a beautiful place. It had two runways and the farm with the crops there. On a summer day when the crops were almost ready for picking and you could see the barley swaying in the wind, you could just close your eyes and picture the B-17s taxiing in. Absolutely fabulous place. We were so sorry to leave it."

Never a group to miss an opportunity: the deconstruction of the Vampire for the imminent Langford eviction meant it was free to travel to Clotworthy house for their VE Day + 60 years exhibition.

Ordnance Survey publications: "I had a little bit of desktop experience."

He was too modest, and whatever extra know-how he needed was provided on the job. He is still there, having guided the magazine masterfully for 13 years and supervising some serious improvements.

The committee did not want the eviction in 2005 to dominate the year entirely and proceeded to plan educational events as it had in years past, though on an understandably smaller scale. VE Day + 60 years was marked by a series of events and displays in Antrim's Clotworthy Arts Centre and the Lisburn Linen Museum. That gave the members something positive and appeared to improve Ernie's disposition.

"The Society's name is being kept in the public eye," he noted at the time.

By the summer of 2005, negotiations for a hangar at the Maze/Long Kesh site were positive enough to begin transporting the collection from Langford Lodge. Ernie Cromie, perhaps a bit too eagerly, gave the go-ahead. Tony Whitehead's group, aware of the Society's October eviction deadline, appears to have made a silent concession for site occupation even while work on the larger Maze Prison master plan was in progress. The deal with the Society was kept secret from the membership as well as the public at that point, even

though any casual observer could spot repeated journeys by large aircraft aboard large lorries heading along country roads in the direction of Lisburn. Government thinking was basic: The Maze prison site was politically fragile, with opposing parties primed to pounce on the slightest appearance of favouritism in the planning process or the final report, due in a few weeks. The secrecy and the delicate dealings with the Strategic Investment Board were the first evidence that the Society's relationship with government—in effect, the new landlords—in the years to come would bring new challenges, and possibly political interference.

Meanwhile, the packing continued at Langford Lodge, well organised but a bit frenzied at times. The same enthusiasm and commitment which had taken the heritage team to Langford were repeated, even excelled, as the members prepared the collection for departure. The Tucano, Jet Provost and Wildcat—at least, the major parts of them—were gone by the October eviction limit, along with vanloads of smaller pieces, display items and artefacts. Four aircraft were still at Langford but, as Ray noted at the time, "the fact that the move had started in earnest helped satisfy our landlords that we were serious."

By this stage, he had put well behind him the shock of being kicked out of the home they had done so much to improve.

If the public could no longer come to the collection, then the collection could be brought to them. At Clotworthy House John Blair showcases his masterful restoration-in-progress of a Nash & Thompson FN.4A aerial gun turret.

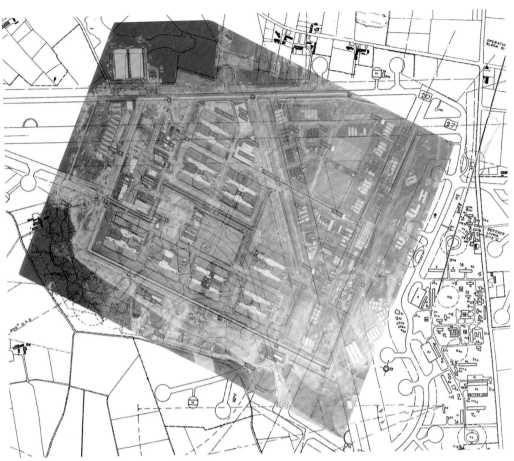

Better known as the site of the infamous Maze prison, this was once RAF Long Kesh. The prison's H-blocks dominate this aerial view, but note the two Second World War era hangars at the top left.

A lot of the hard work—especially on the aircraft and displays—had not been wasted, he said.

"Here (was) a new start, a new beginning, let's get going and make it the best we can," he added, looking at the impressive hangar sight around him 13 years later. "The hangar was in reasonably good condition. There was no asbestos roof, there was clear floor space and we thought, 'Wow! We can actually set our six or eight aircraft here and display them.'"

By the autumn, the only

There would be no airlifts or fly-ins this time, but the transport solution for the Sea Hawk was nonetheless novel: it was towed the 25 miles to the Maze/Long Kesh on its own wheels.

aircraft remaining at Langford were the 'big four;': Buccaneer, Sea Hawk, Wessex and SD3-30. Unfortunately, the inventive techniques employed in the past would not work again. For a start, the Buccaneer was decommissioned so a flight to the new venue was out of the question. And anyhow, there was no runway. As for airlifting the SD3-30, the lowest insurance quote the committee could get was $200,000 U.S—about 200 times more than the best airlift quote in 1993! The bottom line in 2005/06: The aircraft would all have to be moved by road. Another extension to the eviction notice allowed the time needed to plan the last stages of the departure from Langford Lodge.

On December 2nd, the Sea Hawk left Langford for good, taking two hours of towing to travel the 25 miles to the Maze/Long Kesh site. Still, the work teams didn't stop. Working through the Christmas season, they had the SD3-30 and the 15-ton Buccaneer ready in six weeks. They left Langford on January 21st, never to return. The greatest crisis in the Society's 37-year history was behind. Discussions on the terms of its new occupancy were in the final stages and to the group's leaders nothing seemed insurmountable.

One lorry after another passed under the security barrier, continuing gingerly down the disused track past the prison wall. It was an eerie feeling driving through the site, but the working teams shared a couple of things with the eccentric group which, in 1968, had gathered at Aldergrove's Gate 7: passion and potential. Realizing that passion, however, was something else. By February, 2006, the move was completed, leaving the volunteers exhausted, finances drained and the membership down to 270—fallout, largely, from the Langford eviction. The future was vague, not helped by the state of their new digs.

There were two hangars, but negotiations dealt with only

Though their new home remained a closely guarded secret, by the summer of 2005 the smaller airframes had begun to appear at the Maze/Long Kesh.

After years refurbishing Langford Lodge, the members were back to the drawing board with their new home. But it was a home, and a welcome one at that.

Leaving Langford

Hearts were heavy as the heritage team readied themselves to leave Langford Lodge for the last time. The letter below arrived in the midst of the mêlée. It was at once an uplifting accolade for the Society's work and a sore reminder of what was being left behind. The future, however, looked very promising.

Saturday, 18 September 2004.

A long and tiring (but very rewarding) day.

It really is a long way to Tipperary—especially if you drive from south Tipperary to Langford Lodge near Belfast and back on the same day. 437 miles, actually, for the round trip.

Following a suggestion from Rod B. in Adelaide I contacted the Ulster Aviation Society about their museum at Langford Lodge. I received a reply from Michael McBurney at the museum, with detailed instructions on how to find the place

Langford Lodge is an example of how a museum should be run. I felt that I was a friend of the family instead of just a paying visitor, as in some places that I have been to. With the assistance of various members of the staff…I was able to photograph everything that I wanted to. It is definitely a museum that I will visit again. When and where is another matter. Soon after my visit (but not because of it!) the museum had to close. It seems that the owners wanted their hangar back.

Peter O'Connell

one, and it lacked running water and electric power. The buildings were dilapidated and, significantly, public access to the site was heavily restricted. How could anyone even approach the property without the feeling that perhaps they weren't welcome?

Still, the heritage team members were just happy that the move was behind them and they had a roof above.

"My overriding emotion is one of thankfulness—for the 13 years we had to build the collection and to learn some useful lessons about running an aviation heritage museum," wrote Ernie Cromie in the March, 2006 *Airmail*. His sense of relief was palpable, and with a major and very positive planning report now in the public eye, it seemed the future was not just bright, but dazzling.

One lorry after another passed under the security barrier, continuing gingerly down the disused track.

The biggest aircraft were the most challenging, but thanks to W. J. Law Plant & Transport the Society once again had professionals and supporters on hand to guide them in the complex move.

The move to the Maze/Long Kesh was bittersweet. The home they had built at Langford Lodge was gone forever, but a new chapter in the UAS story lay ahead, and it was to be the most exciting chapter yet.

4.1 | Ground Bound Adventure

There's something primal about an air show. It's all very basic, but with a frisson of excitement. A warm day, the milling crowds, the frigid bite of an ice cream cornet and the greasy fragrance of sizzling burgers. And—oh yes—a frenzy of aeroplanes snarling above like pturbo-pterodactyls.

It comes to that: Up in the sky, where we wish we could fly. And from time to time, we shake our heads in wonder at the aerial tricks those pilots perform.

But there's also a dose of wonder down below, teasing visitors to see, touch and marvel at the very machines that have woven their own magic spells above and through the years.

Welcome to the Ulster Aviation Society ground display!

It might be at Enniskillen, Antrim, Newcastle or a dozen other festivals in a given year.

Since the early 1980s, the Society has attended nationwide events as participants, not just watchers. The Langford exodus put an end to open days until the collection was well established at its Maze/Long Kesh compound.

That was near the time, too, when the Society's leadership decided to bring its aircraft to community events.

The payoff in terms of public enjoyment was tremendous, but the logistics were a challenge and the sheer workload demanding.

It is early morning on Friday, September 1st, 2017. The events team has gathered at the Maze/Long Kesh hangars, getting ready to pack up and head for Portrush, 65 miles away. The Air Waves show begins on Saturday afternoon.

This year's earthbound effort will include four aircraft: Alouette and Scout helicopters, the Ferguson Flyer and the replica Spitfire. The volunteers will also load the Society's survival equipment display (complete with ejector seat), the turret, the shop items and interpretive displays.

Aircraft are strapped onto low loaders and trailers—a bit of a riggers' ballet—with smaller bits crammed into the Society van. The first convoy heads north. While the next load is prepared at the hangars, the first team is busy at Portrush.

By late afternoon, the second load has arrived and the setup drill is in high gear: Helicopters are in place and aeroplanes nearly assembled. The Spitfire team always has a tough job, well-practised though they are. The wings are heavy and ponderous. Each move must be done efficiently and safely, with a couple of members ensuring casual visitors stay well back. Tomorrow the public can get as close as they like. Mal Deeley sets up the survival display, John Weir and Ray Spence have the shop almost ready while Fred Jennings organises the books. Most of the non-aircraft displays are in a large marquee tent loaned by the local council. By twilight, everything is in place and the exhausted group heads for dinner and a well-deserved sleep. Saturday will be another early start.

By 9 a.m. the setup team—now the core of an event staffing crew—has been drifting over to the Society's site.

Air Waves Portrush is Northern Ireland's biggest air show. Two days, 200,000 visitors, 18 aerial displays and all sorts of ground displays and exhibits. For the Society, it's the greatest outdoor effort of the festival season. Fortunately, help is funneling in as other Society members arrive to assist the main events team. The air show brings the membership out of the woodwork like no other event. The award for distance travelled goes to David Mellon, who has journeyed all the way from his home in Dublin. There's only one little hiccough; the small Society tent has blown over in a gust during the night. With quick action and a bit of duct tape, it's put right as the first visitors begin to arrive.

The official start to the day's proceedings is 11 a.m.; time enough beforehand for the volunteers to grab a coffee and tweak the last-

It's all hands on deck as the Society's biggest roadshow makes its way to the picturesque north coast.

minute setup details. Their experience in recent years means they know what to do without a formal plan. Each exhibit has—more or less—a supervisor and the team members rotate their duties so that everyone gets a break from time to time.

Well before the aerial displays, the crowds are arriving, picking out their ideal places for viewing the show or simply wandering around the many displays, queueing up for the children's rides and games, grabbing a snack or chatting with friends they've met. At the Society location, volunteers are meeting visitors, shepherding children into the Spitfire or Alouette cockpits and explaining the displays. New for 2017 is member William McMinn. The intrepid pilot of the Ferguson, who flew it briefly the year before, delivers a firsthand account of what it's like to fly a Harry Ferguson design from the dawn of flight. It's the only aircraft there which the public cannot board; it's too fragile and very valuable. However, the visitors are encouraged to try out the cockpits of the other machines—and they do, in droves. The replica Spitfire, with Fred Jennings, Harry Munn or John Martin as stewards in place, is especially popular, with the Alouette a close second. The chatter is lively and interested:

"I didn't even know there was a collection! Where do you have it?"

"This helicopter crashed in a Donegal lake in 1985. The Irish Air Corps donated it to us."

"Our Spitfire? It's a copy of one that flew here during the Second World War. The original was one of 17 bought by the public in a *Belfast Telegraph* campaign."

"Yes, this rickety, fabric-covered machine actually flew!"

Around 1 p.m., as many families are tucking into their picnic lunches, the flying display starts with a bare whisper: a graceful aerobatic set by the Ulster Gliding Club, complete with a beach landing right in front of the crowd.

It's followed by a screeching contrast as an RAF Typhoon jet fighter slices the sky and booms upward at a steep angle before banking right and returning. The crowd roars its pleasure.

With visitors distracted, the Society volunteers grab a chance to relax and enjoy the show. Kathy Scott has brought boxes of

The one-of-a-kind Ferguson Flyer is a special attraction for visitors to Air Waves Portrush.

sandwiches, young member Ben Gibson arrives and spells off an Alouette steward. Jim Robinson pops into the Society van for a five-minute kip.

Typhoon display over, the crowds rush back to the Society's site. Calendars, T-shirts, hats, books and scale model kits fly into eager hands, courtesy of a formidable sales team eager to please before the next showpiece in the aerial running order. Donation buckets rattle and the Society camaraderie hits a new high. Events like this make all the effort worthwhile.

The Red Arrows aerobatic team closes the show with its ever-popular routine, but the end of the flying means even more free time for the crowd to visit the Society's site. The volunteers are busy right until early evening.

Sunday will bring more of the same, weather permitting. It will end with the arduous task of dismantling the displays and tables, re-packing the few remaining sales items and loading the aircraft aboard the Walcon transport and our trailer.

Next stop: The Society's hangars at Maze/Long Kesh.

The teams and the leadership will assess the results in the days to come, but it's clear that Air Waves 2017 is as big an event as they could probably manage. Maybe not, says John Weir, salesman extraordinaire. How about bringing the huge Puma helicopter? Hmmmm….maybe next year.

Pilot William McMinn stands astride the Ferguson. He's the only man alive who could give visitors a first-hand account of flying the delicate craft.

After enjoying the aircraft displays, many visitors pop by the Society shop looking for a souvenir of a grand day out.

C H A P T E R F I V E

A lot of people are dreamers. Look around. The leaders of the Ulster Aviation Society, over a 50-year period, have been among them. During the winter of 2005-2006, two groups of dreamers were nursing hopes for their future at the prison edge of Lisburn.

The aspirations of one, the Ulster Aviation Society, were built on enthusiasm and hard work, tempered by the trying experience of a gypsy existence since its formation. The volunteers' aim was simple: They would fashion a hangar into a home for the Society's collection. The second group was a design team from EDAW, landscape architects with a major world reputation. Their dream was much more ambitious but no less admirable. The Northern Ireland government had commissioned that company to design a master plan for the entire Maze/Long Kesh property. But EDAW's dream would eventually quake, then crumble under the weight of relentless political bickering.

The two heritage hangars date back to the Second World War. Only the far building was in use by the Society at the time of this photo in about 2010. The near building was used for government storage.

Its master plan, revealed in January, was grand: a 42,500-seat sports stadium and a "conflict transformation centre" would be the showpieces. Also proposed were a 5,000-seat indoor arena, a hotel, office buildings, cafés and a multi-screen cinema, plus housing for hundreds. The hangar collection of the Ulster Aviation Society would be a part of the larger scheme. But proposals for the conflict transformation

> "It felt like an historical moment, that we were looking back in time."

center became embroiled themselves in months, then years of disagreement among politicians, effectively ending or stalling any development at all. And that, in turn, was evidence which foreshadowed the eventual disintegration of power-sharing at Stormont and the Assembly's collapse in 2017.

But the Society forged ahead, unencumbered by political divisions. The leadership, too, believed in something truly ambitious, albeit on a smaller scale. Volunteers quickly found their niche at their new location—and occasionally a surprise, as John Martin recalled, when working at the northwest corner of Hangar One.

"We got permission to take down the concrete blocks that had been put up to close off certain doors in some areas of the hangar. So we hired a Kango hammer for the weekend and… just started drilling out the blocks."

They discovered that the original builders had used high-strength concrete blocks.

"So it made the whole job incredibly difficult," he said. "We were all standing there until our arms just couldn't hold the Kango hammer up any more."

But once they'd punched through, they found they had hammered their way into history. They were in the old wartime kitchen, with canteen hatches, plumbing and such.

"It felt like an historical moment, that we were looking back in time when the hangars were used for their original purpose. It just brought it all back, you know?"

On the other side of the hangar, Paddy Malone and Harry Martin picked out three small rooms in an annex connecting the two hangars which would provide perfect space for exhibitions on the Royal Observer Corps, a subject of considerable passion for both. They picked up their tools and bent into a new challenge.

Other volunteers, like Joe Fairley, got stuck back into aircraft restoration. He set about straightening the twisted

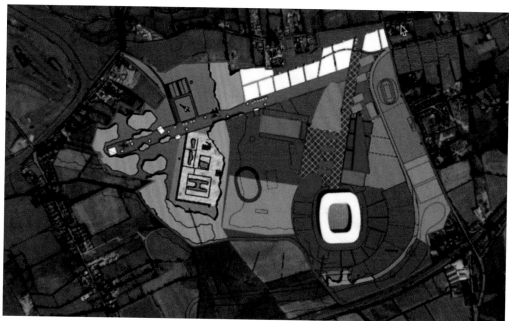

A layout from the master plan of 2006 shows a large oval stadium (lower right) and an expanded area for the Society's collection (upper left) beside its two grey hangars. Note the little aircraft, with one the shape of a Concorde!

A volunteer inspects an annex room. It took years to put the buildings into half-decent shape after the last aircraft had arrived from Langford Lodge in 2006.

propeller of the Wildcat, the force of its crash landing being enough to warp the metal blades.

Others began the painstaking reassembly of bulkier aircraft which had been dismantled for their pilgrimage to the Maze/Long Kesh site. A series of break-ins had spoiled the serenity of the Christmas season, doing little to lift the spirits of the weary workers. But with a stubbornness now typical of the Society, they soldiered on.

Amidst all of this commotion was a distinct common goal: Put the Society back on the public radar.

Step One: Stake your claim. The management committee wanted a museum in the making and, eventually, something beyond a mere collection of aircraft and other items. A successful application to the Northern Ireland Museums Council in February of 2007 made that clear. The Society would want, at some point, to be seen as an organisation that researched, developed and explained its collection in an attractive, fully accessible setting and in a manner central to Northern Ireland's aviation heritage.

Step Two: Demonstrate potential. The public was not allowed to access the Maze/ Long Kesh site at this stage. Even admission for Society members was limited by the site managers to those dedicated few who volunteered week in, week out to transform the husk of a hangar into a magnificent museum. Ultimately, the

Another annex room to challenge Society volunteers. Even ten years later, junk like this nestled next to aviation artefacts awaiting display space and the members' time to restore the rooms and arrange the treasures.

The award ceremony was a strong endorsement of the Society. Since then, a wide variety of groups have shown up, fascinated by the venue.

access decision rested with Northern Ireland's politicians— specifically the Office of the First Minister and deputy First Minister (OFMDFM).

An exception was granted in 2007 to allow the hangars to become the venue of the annual McKillop Irish Spirit award, granted to the Rev. Peter McVerry for his work with the homeless youth of Dublin. Society members basked gleefully in the opportunity to open the hangar doors for the first time to outsiders.

Step Three: Educate. The

volunteers had revelled at Langford Lodge in sharing their collection and their passion with anyone and everyone who showed an interest. But with visits to the hangar out of the question for the moment, how could the public be reached? Society secretary Paul McMaster and his cameraman brother had one answer: film. With the help of producers Subterranean Fish as well as a grant from the Heritage Lottery Fund, a professional production was in hand. The subject was Northern Ireland's part in the air campaign during the Second World War. It was geared toward the secondary school curriculum.

Northern Ireland's Battle of the Skies was an able demonstration of the Society's ability to produce professional educational material. Even more significant, it exposed the wealth of knowledge within the collective membership. The film, on a new medium—digital video disks (DVDs)—was presented to 300 grateful secondary schools on a complimentary basis. Screenings of the production remain a staple of class tours to the Society collection today. Producing an agreement with government bureaucracy proved to be a totally different matter.

Like the film production, the move to Long Kesh—as challenging as it was—had nonetheless been done with relative expediency. The Society's leaders discovered that government officialdom rarely worked like that. It would be January, 2009 before a formal licence was agreed for the Society's occupancy of what was now known as Hangar One. And even once it was signed, the terms were restrictive. They allowed for pre-arranged hangar tours by the public, yet the Society was not permitted to advertise the tours on traditional media or online. Word of mouth could be the only recourse.

It was a huge blow to the Society. It was bound not just constitutionally but passionately to the development of a museum of aviation—which obviously suggests a direct connection with the public. Still, a formal licence was a start, and hopes

in those early years were high for the future of the Maze/Long Kesh site. The restrictions might be only temporary.

Anyway, each Society member—and most were not hangar volunteers—had something more than mere access to a small but unique aircraft collection. For their £17 annual membership fee, they also had a subscription to a monthly magazine, an occasional chance at an interesting flight or two, and monthly meetings which featured a wide variety of speakers who were authorities on specialised branches of aviation. It was evidence of the continuing, resolute efforts of the management committee that in the turmoil of relocation members were not left wanting.

Even back in 2008, while uncertainty prevailed over a licence agreement, the Society's leaders were determined to mark the 40th anniversary of the organisation in a special way. Forty years of dedication, challenge and persistence? Surely 2008 was a vintage year! So Eric Gray and Ernie Cromie proposed to mark it with flights in a vintage aircraft. They announced that members could purchase tickets for one of a series of farewell flights of Air Atlantique's DC-3, that venerable transport queen, itself a design more than 70 years old. Members leapt at the chance and tickets sold out in record time.

It was an opportune time as well to thank Society patron Sir Philip Foreman, with a tribute book by historian Guy Warner. *Shorts—The*

A delighted Paul McMaster jumped at the opportunity to fly aboard a DC-3. By 2008 rising costs and diminishing airframes meant vintage flights were a rare treat.

The Society's first foray into digital video was aimed squarely at the education sector, specifically secondary school children.

Foreman Years was launched in March, 2008 at Belfast Central Library. The venue included a Society photo display enjoyed by a receptive crowd which included the great man himself.

By mid-2008 the heritage team had put the upheaval of the move to Long Kesh behind them. They gaped at the cavernous space, devoid of upright pillars and low-slung rafters. They were keen to fill it. That could begin now, because things had

The Shorts SB–4 Sherpa is unique in the world, but the missing wings posed a major challenge for Society craftsmen.

also begun to change within the aircraft collection itself.

The committee had put the message out: The Ulster Aviation Society was back in business and looking for more aircraft to join their growing assemblage. The first response was positive. The Imperial War Museum was seeking a home for its hand-me-down Shorts SB.4 Sherpa, designed and built in Belfast. The revolutionary aircraft—minus its wings, tested to destruction and lost since its heyday in the 1950s—arrived on July 17th, 2008, to join its Shorts siblings, the SD3-30 and Tucano. Restoration would have to wait a bit while work proceeded on the Wildcat and other subjects.

The committee had put the message out: The Ulster Aviation Society was back in business.

The big SD3-30, for example, sat half naked with no wings. They'd been detached for the move from Langford Lodge. Work began that same July— and ended almost immediately, to the relief of Ray Burrows.

"Having readied ourselves for the re-installation of the Shorts SD3-30 centre-wing section, we achieved it in less than half an hour!" he boasted. "Impressed? I certainly was, but even more so whenever we attached both outer wings in less than two hours."

Through more demanding experience—especially with the Wildcat—the heritage group knew that future restorations would likely not be so simple.

For the moment, though, they turned their attention to another heritage matter. The coming year, 2009, would mark the centenary of powered flight in Ireland. Make no mistake: Many of the volunteers felt as strong an attachment to aviation history as they did to the challenge of mating wing to fuselage or removing rusty nut from stubborn bolt.

Young Harry Ferguson fitted that latter category. He had charted his course into aviation history when he coaxed his monoplane into the air on December 31st, 1909. Although he traded aviation for agriculture in later years, the County Down man's place is secure as the first to design, build and fly a powered aeroplane in Ireland.

A series of events and commemorations were planned for the centenary of that event. In preparation, Ballynahinch octogenarian Harry Pickering got to work refurbishing the 1981, one-third scale, radio-controlled replica of the Ferguson monoplane which he had loaned to the Society some ten years earlier.

The model's first outing was fittingly a 'Picnic in the Park' in July at Hillsborough. It drew much attention from some puzzled onlookers—a reaction probably familiar to Ferguson himself many years earlier. Ernie Cromie delighted in celebrating the achievements of Ferguson while fulfilling his Society's remit of education: "Interest in the model was very marked and timely, to judge by the frequently-heard comment, 'I never knew that Harry

Ferguson was a pilot.' Need I say more?"

Further tributes soon followed: an obelisk in the same park bearing a commemorative plaque designed by renowned sculptor John Sherlock, a lecture with Guy Warner and Ernie Cromie at the Island Centre in Lisburn on August 20th and, two days later, the Society's own Ferguson-themed open day for the public at the hangar. It was the first open day at the site.

For a time, it looked unlikely to happen at all, with permission from the Office of the First Minister and deputy First Minister arriving a mere five weeks before the open day was scheduled to take place. Though only a nuisance at the time, the Society didn't know that such delays—and worse—would be common in the years to come. Still, the day's organisers weren't going to let the short notice that summer hold them back. They proceeded to arrange a stellar event, albeit relatively low key in light of ongoing demolition work at the nearby prison complex.

A crowd of around 1,000 Society members and the public alike descended on the Maze/Long Kesh site on August 22nd. The Harry Ferguson Celebration Committee brought some vintage tractors and also made a surprise presentation to the Society of a painting by Gerald Law, himself a Society member. The art depicted young Harry's momentary departure from the earth which sustained celebrations a whole century later. The visitors swarmed the stalls and marvelled at

the collection's aircraft to a drumming soundtrack of no fewer than 16 visiting helicopters which came and went through the day. There was a real buzz about the hangar. It was clear evidence of the Society's true potential should the restrictions to their tenure ever be relaxed.

Once the crowd dispersed and the clean-up was over, the Society's volunteers found some time—not a lot—to consider how the hangar collection might look with new exhibits to manage.

There was talk of a Canberra bomber or reconnaissance aircraft perhaps becoming available, but most of the hangar team was busy with work already at hand. As well, there was much to be done with upkeep of the hangar itself and the development of museum quality displays. There were new exhibits to manage including examples of rocket weaponry: a full-scale replica V-1 (Fieseler Fi.103) flying bomb built by Society member James Herron and a Shorts Seacat missile system. An unusual homebuilt aircraft, the parasol-winged Clutton-Tabenor "FRED" (G-BNZR), was donated by local man Mervyn Waugh. These and certain other donations occasionally raised the question of what specific significance they bore to Northern Ireland aviation, but the Society wasn't overly particular at the time. After all, said some members, certain major aviation museums in other countries weren't so discriminating, so why should the Society be different?

Engineering as artwork: The turbine blade assembly from one of the Society's many aircraft engines.

Still, as welcome as they were, the rate of new arrivals compounded the pressure on the volunteer workers. There sat the De Havilland Vampire, for example, five years since its move to Maze/Long Kesh and still in pieces, the victim of higher restoration priorities.

The members were hardly overburdened. A point of pride for the Society's crews at the scene was that all the work was wholly voluntary. If a worker was fed up, he could walk away and tackle the problem another day. If a task drained the brain or the energy, it could be sidelined until someone with renewed stamina or the right skill set came along. Simply put, the volunteers loved their work. And as days became months and the progress in Hangar One became obvious, they enjoyed it even more. The spacious interior slowly improved, from a dumping ground for aircraft husks into the semblance of a museum. The hangars in their less-than-adequate state nonetheless had a character of their own. As members painted walls, mopped floors and tidied rooms, a camaraderie unique to the Ulster Aviation Society blossomed. The numbers grew as well. By 2010, any given Saturday would see 20 regulars in attendance, with the total often pushing 30. The buzz of the hangar community remains a draw for new volunteers to this day, and only heightens with each new recruit.

By this stage of its history, the Ulster Aviation Society may have appeared—by virtue of its dedication to advancing

Gerald Law, Society member and talented artist, portrayed the first exploits of pioneer Harry Ferguson in 1909.

Gerald.W.Law

its aircraft collection—to be developing solely an oversized heritage project. That perception would be flawed. The Society's leadership was keen to make the organisation's presence felt in other areas as well.

Harry Ferguson's flying feats of a century before, for example, still loomed large in the public's mind, so the Society made the most of that.

In August, it launched a three-week exhibition in the Newcastle library dedicated to Harry Ferguson. That initiative actually followed a series of events including a parade through Newcastle town centre of the one-third scale monoplane replica. The cue for that occasion was the 100th anniversary of his award-winning aerial jaunt over Newcastle beach in August, 1910. One happy entry in the exhibit's visitor book exulted: "If Harry Ferguson had seen what Newcastle did for him over the weekend, he would have been very proud, I'm sure." There was pride reflected as well in other centenary events honouring someone whose aviation success was arguably even greater. In August, 1910, Carnmoney's Lilian Bland took to the sky in an aeroplane she designed and built herself—the first woman in the world to do so. With a touch of whimsy, she had christened her machine the 'Mayfly,' (as in 'It may fly, but it may not'). And fly it did, vaulting Bland into the air and into the history books in one deft swoop. The Society organised a wreath-laying ceremony at the great woman's grave in Sennen,

Many examples of aviation ephemera such as this are part of the Society's collection.

The distinctive but dismembered Vampire had to wait for years before volunteers could find time to nurse it back to display condition.

The team spirit in the hangar was palpable. Harry Munn (left) and Jim Robinson were a double act to rival any creative pair.

Cornwall, as pilot Mark Hillier of a local flight training school flew a salute overhead. Guy Warner also did his bit to mark the centenary, with the publication in December of his biography about the tumultuous life of a woman who was Bland by name but not by nature.

The unpaid nature of the members' work meant they clocked no hours or progress. They were left to work quietly at their own pace. The time spent on any one project could no more be recorded than could their dedication and passion.

Those factors caused the septuagenarian Eddie Franklin to collect his jaw from the hangar floor on March 30th, 2011.

No longer a mere collective of enthusiasts, the UAS had become a true community. They regularly downed tools to celebrate acquisitions, birthdays and, despite hangar temperatures, Christmas.

Society members were eager to spread the word outside the base of their collection. The Newcastle parade in 2010 was an ideal education opportunity to rekindle the town's relationship with Harry Ferguson from a century before.

Ulster Eyes in Global Skies

The largest aircraft in the collection of the Ulster Aviation Society looks pristine and powerful now, a mirror of its once-familiar form in the skies over at least four continents during the Cold War.

All 23 Canberra PR.9s, including the Society's XH131, were built at Shorts' Belfast factory, from where they rolled out in equally immaculate condition. They headed off into operations over the years over such varied locations as Indonesia, Afghanistan, the Middle East, the Balkans, Chile and—yes—Northern Ireland. Canberra XH131 played its part in many of those missions.

Since its 2006 retirement, however, it had lain dormant and rusting on a grimy hardstand of a storage area on the former RAF airfield at Kemble in Gloucestershire.

Its private owner, who had once offered the old reconnaissance aircraft for free, kicked the price up to £10,000 in 2010 when a Malta museum also showed interest.

The Society had no money like that to fork out at short notice. However, secretary Paul McMaster provided a quick loan to complete the purchase. The Heritage Lottery Fund also provided a £50,000 grant to fund the dismantling, shipping, transport and education materials.

The Society contracted a firm in England to dismantle the big bird, and that's when the problems began. The tale of woe is long and complex but the bottom line was simple: The Society paid thousands for work that wasn't done as promised.

Ray Burrows, who had led the charge to get the Canberra, recalled the guarantee from the man assigned to XH131: she would be on the Liverpool ferry on June 7th, 2010. Ray phoned him two days beforehand and remembers his own astonishment.

"'Well,' he said, 'I don't have the engines out yet and I don't have the wings off yet.' And I dropped the phone. I didn't even talk to him. I just dropped the phone...I told the committee, 'This guy has taken us for a real ride.'"

Ray, David Jackson and Mal Deeley—none of them with experience in dismantling Canberras—headed for England. During several visits over a period of months they gently (usually) pulled it apart, with help from volunteers from near and far, including a crew of four from the South Yorkshire Aircraft Museum at Doncaster.

Another key figure was Steve McMaster of Cobham Engineering at Poole, who arrived to confront the problem of eight rusted and stubborn wing bolts.

The first of those bolts was already out, after eight hours of steady and frustrating persuasion by the Society's crew.

"What you need," said Steve, "is a hydraulic extractor."

Right. One of those wee jobs which can provide a force of 20 tons. "It was slightly bigger than a baked bean tin and cost £400, but each bolt came out in less than five minutes," said Ray.

During weeks of effort through the summer and into the winter, his *Ulster Airmail* stories kept the Society's members apprised with progress on the project. His final report covered the night when the weary troupe of volunteers arrived back at the Maze/Long Kesh hangars just before Christmas, 2010 with their precious cargo in tow.

Chairman Cromie's words of praise afterward said it all: "It is nothing short of a triumph and demonstrates...just what can be achieved by well-led voluntary initiative and enthusiasm in pursuit of a worthwhile goal."

For a detailed history of the big reconnaissance aircraft, see The Last Canberra: PR.9 XH131 *by the Society's Guy Warner, available at the UAS hangars.*

Based on a 1950s design, the Ulster-built PR.9 was the ultimate development of the Canberra. XH131 served the Royal Air Force for 46 years until its eventual retirement in 2006.

In this twilight photo at RAF Kemble (Cotswold), Society volunteer David Jackson cuts a lonely figure aboard the fuselage of XH131. It's early in the recovery process, with much work ahead of the team.

It's a moment of quiet pride as the right wing is lifted away from the fuselage. The attachment bolts for the left wing proved to be firmly fixed, but the team soon learned how to remove them.

Another impressive tenant arrives at the Society's collection hangar. It's 2010, and the challenge of re-assembling XH131 lies ahead.

The Society's triumphant team of (left to right) Mal Deeley, David Jackson and Ray Burrows. Not present for the photo were the many helpers who arrived at the Kemble site at various times to assist in the work and the transport efforts which eventually brought the Canberra to the Maze/Long Kesh hangars.

It lacked Langford Lodge's runways, but Maze/Long Kesh provided ample space for rotorcraft. The Irish Coast Guard were keen to support the Society's re-launched open days in 2011.

His visit was a welcome surprise for protégé and old friend Ray Burrows who welcomed his mentor back into the fold. His visit was timely. Ray had authored a series of articles about the history of the Society up to that point, and Eddie was the ideal man to introduce it: "In the beginning, God created anoraks and I was one of them!" Another legacy of Eddie, *Ulster Airmail*, continued to prosper—and to travel, perhaps even to Buckingham Palace, to hear Ernie Cromie tell it.

The time spent on any one project could no more be recorded than could their dedication and passion.

He was curious one day that same March about a throng gathered outside Hillsborough Castle near his home in the village. Snatching a batch of freshly-printed *Airmails*, he elbowed through the crowd just as Prince William and his fiancé Catherine approached. A delighted Ernie thrust a copy of the magazine forward in hope that the prince, a former RAF pilot, might notice the aircraft on the cover. He did, and after a brief chat he departed the town with a

Bland by name and bold by reputation, Lilian is rarely seen as the beauty depicted here. Adventure was her middle name.

copy of *Ulster Airmail* in hand. The relative quiet of 2011 enabled the committee to plan another UAS open day, having missed the opportunity in 2010. June 11th was the date. The 70th anniversary of the Blitz was the theme. The local Air Training Corps (ATC) cadets volunteered to marshal the event, leaving Society members to meet, greet and entertain the public with tours of their exhibits, to the music of the Swing Gals. The quartet took their audience back to 1941 with a repertoire of vintage songs from wartime U.K.

The event was a success on a minor scale. Part of the problem was the recurring delay in obtaining permission each year from OFMDFM. The best open days require long-term planning, a point apparently lost on the politicians and functionaries who considered the annual application and often responded at very short notice. With events on the site totally in government hands, there was little the committee could do. The 2011 episode prompted Ernie to quote Henry Ford: "When everything seems to be going against you, remember that the airplane takes off against the wind, not with it."

Ernie, a skilled diplomat, well organised and patient, had flown against the wind from time to time during 30 years as chairman, but by 2012 he was inclined to glide into less onerous activities in the Society. He had been a strong but humble leader, never too proud to mop the hangar floors or wipe the counters in the crew room.

He had met difficult challenges, guiding the Society through the failure of the Newtownards museum project and the eviction from Langford. He had been active in the heritage group, helping increase the collection by one aircraft after another. Ray Burrows, his strong partner in the heritage efforts, was the obvious candidate to assume the role as chairman. At the annual general meeting in March, he paid due credit and thanks to his predecessor: "The Society [was] on the brink of fulfilling a dream we had years ago, of the best aviation collection of its kind in Ireland." And that dream was fulfilled. Looking back years later, he told Ernie: "We'll never be able to repay the debt we owe to you."

Speaking for himself in his final column in the *Airmail*, Ernie said he left with mixed feelings but, he added, "my overwhelming emotion is one of thankfulness." Among his many successes in the organisation, he said the adoption of a constitution and its acceptance as a registered charity stood out as his personal highlights.

Ray had been a passionate Society member and volunteer from its earliest years and served in various management committee positions as well. His first days as chairman oversaw the aircraft arrivals continue at an unprecedented pace. And as hangar boss, he was immersed in organising restoration crews and priorities as well as getting involved himself.

The hangar volunteers, numbering about 40 at this

Most enthusiasts have a favourite aircraft or three. Ernie Cromie's top choice was the P-38 Lightning, hundreds of which made Langford Lodge their home in wartime. Ray Burrows (right) presented this painting to mark Ernie's retirement in 2012 from the chair of the Society.

"My overwhelming emotion is one of thankfulness."

time, were never short of work.

Teddy Colligan got stuck into the Gannet fuselage as David Jackson and Joe Fairley attempted to rebuild the severed propeller blades. The endless Wildcat restoration continued with perhaps the biggest team it had ever seen, among them Alan Moller, Terry Neill, Harry Munn, Jim Robinson, Stephen Riley, Wolfgang Wenger and Ray Burrows. Leonard Craig installed fresh ejector seats in the Jet Provost as Fred May, son John and Jim McCartney constructed brand new components for the Sherpa wing. Other volunteers cleaned and painted the building, hauled

Ray was introduced to the Society by Eddie Franklin himself. In 2012 he took charge of a dynamic community utterly transformed from that fledgling spotters' group of the late 1960s. Here, in 2011, he prepares the Canberra for its dedication event.

parts, built jigs, replaced worn wiring, prepared graphic stands and so on. From one end to the other the hangar was alive.

An opportunity to show off all this work came around once again on August 25th as the UAS held another, now annual, open day. By this time in its history, the leadership of the Ulster Aviation Society had the experience and keen awareness to how important such occasions were. Certainly, regular open days could provide the income necessary to sustain a museum in the making, and to pay for further acquisitions and expansion. But there was more to it than that. Open days provided a direct connection with the public. An open hangar door became an entry into a fascinating part of a visitor's heritage—one that most of them were probably not aware of. The Society's collection was a visible story, and as such

Undercarriage indicators from the Wildcat, the rusted one beyond recovery. Hangar visitors can witness restorations in progress.

it would thrill the visitors. If it was well told through objects and personal connection, they wouldn't forget how the experience made them feel. The challenge to Ray Burrows and his open days team was to ensure beforehand that the event would be positive. That meant weeks of planning and preparation for the biggest event the Society had ever held—not just a showcase, but a festival. The team booked musicians, staging, portable toilets and set aside space for food vans. They arranged for children's activities, including a bouncy castle, and set aside parking for hundreds of cars. They booked helicopters for a fly-in. The "hangar mob" moved aircraft, engines and display cases, dusted them off and tidied the exhibit rooms. A volunteer base of 74 members and friends was organised to ensure it all ran smoothly. The theme this time was a tribute to "The Boys of '42"—

the young American servicemen who had arrived in Northern Ireland some 70 years earlier to do their part in the war effort.

"We're talking about ordinary men," explained Ray Burrows. "Some of them were farmers, some of them were clerks, some of them worked in offices."

A Society video crew filmed the preparations as well as the open day activities for a production which would also reflect the historic American connection. A tractor had towed the Buccaneer outside and it treated appreciative crowds every half hour to a performance of wing folding, wheel lifting and airbrake deployment. Arriving visitors were greeted with a bizarre symphony of whining hydraulics, rotor thumping and crowd applause, all to the accompaniment of the Swing Gals' singing, drifting out from the stage inside the hangar. Among the guests were the lord lieutenant for County Antrim, the United States consul general and two warders from Hillsborough Fort, dressed in period costume.

Poor weather forecasts were unfounded; the day remained dry. It was, in Ray Burrows' eyes, "an absolute success." Attendance was over 2,000— doubled from the restrained open day of the year before.

Of course, the Society provided other opportunities for the public to see the collection. Pre-booked group tours by all sorts of organisations continued, each with its own Society guide. It was an approach which went down well with the visitors, who applauded the personal touch. One such guest, *Aeroplane* editor Jarrod Cotter wrote in the visitor's book: "A fantastic display of aircraft, looked after by the most enthusiastic and dedicated group of volunteers." Schools also continued to avail of a tangible connection to local history, with over 20 class visits in 2012. The students could see history, touch it, even climb inside it, and that was a lesson for the organisers as well as the children.

The tactile experience had long interested Guy Warner as he considered outings for the members. One that stands out was another visit by the biplane *Iolar,* the De Havilland Dragon which had proved such a hit years earlier. On May 19th it took to the air at Newtownards with a load of pleased passengers. It was, in Guy's words, "like stepping back in time to a more elegant, yet simpler age, when flying was an adventure." To date it remains the last vintage aircraft flight arranged by the Society.

As 2012 passed by, the Society members became increasingly focused on the future, especially in regard to communications. Treasurer Garry MacDonald established the Society's Facebook page. The committee, mindful of a growing media interest, established a public

A radial engine gear and collar assembly reveals the art of engineering.

Jennings, installed a working transponder screen, delighting visiting guests who could watch with interest a live schematic of air traffic movements all over Northern Ireland.

In half a century of history, the Ulster Aviation Society could tick off some interesting years. Some were good, some bad, a couple even a bit scary. But if any one year could stand out, 2013 might take the prize. Membership reached a record high; Membership Secretary Des Regan's flock had soared to 360, finally cracking the post-Langford slump. The bank balance was firm as well, erasing a worry that had nagged for years. And, though it seemed no

Working with no blueprints, but a few basic design drawings and models, William Smyth and Jim McCartney (nearest camera) install a new metal spar bracket for the Sherpa's wooden wings.

It's open day 2012 and Ray Burrows briefs the volunteers on their duties. Behind the scenes were days of planning, preparation and practice. This bunch will have a very busy day ahead.

more than just a handy thing to do at the time, the Society got involved in outside community events on an unprecedented scale. Time would show that such involvement would not just be a delightful adventure; it would be necessary to the Society's survival.

The catalyst that kick-started the auspicious year fell right into the Society's lap. Over in Belfast, the King's Hall exhibition centre was showing its age— 117 years at the time. From 1933, it had been the home base for the huge Balmoral Show, annual showpiece of the Royal Ulster Agricultural Society. In 2013, the Maze Long Kesh Development Corporation (MLKDC) sold the RUAS a large corner of the Lisburn site. It could be the first gem in the crown, once development was agreed—if it was agreed. They would soon see.

Pending construction of a huge exhibition centre, the RUAS would store much of its equipment in the "twin" hangar adjacent to that of the Ulster Aviation Society. And, in a generous effort to know its aviation neighbour, the RUAS invited the Society to join the displays at the 2013 Balmoral Show.

An opportunity to interact with tens of thousands of show visitors, with only the width of the Maze site to cross? The committee leapt at the chance.

On May 15th, crowds flooded into the new format Balmoral Show, astonished to see in the midst of it all a partly-restored

Second World War naval fighter aeroplane and its carers, eager to introduce themselves and their cause to each and every passer-by. It didn't hurt, either, that the donation buckets returned to the hangar filled to the brim.

The word began to spread: The Ulster Aviation Society can travel to your show, a unique addition tended by enthusiasts with an intriguing story to tell. There was a festival at Moira and the air show at Newcastle, each one providing opportunities for the team to hand out invitations to their own event: UAS open day 2013 in August.

This time, the site landlords (ultimately, the OFMDFM) had provided ample time to prepare. That was fortunate, because the event would be the biggest in the Society's experience. Ray Burrows had hardly slept the night before, nervous about the weather and a welter of problems he could do little to control at this stage. "The thought of something going wrong, somebody having an accident … having to call an ambulance, that gave me hours and hours of worry." But when the morning dawned, mild and welcoming, he had nothing to fear.

Eager hands had hauled six aircraft from the hangar to the outside compound: Buccaneer, Wessex, Sea Hawk, Alouette, Air and Space 18A gyroplane and the little red Robinson R22. Parked across the road was a cluster of vintage vehicles including a trailered anti-aircraft gun, next to a large space for the helicopter

Ernie in charge of a favourite machine a few hours before the big day. He's been a stickler for keeping the hangar clean and tidy.

You meet all kinds of people on open days. There may be a mannequin hiding in one of the Society's many mixes of outfits.

fly-in. Visitors would even be offered flights around the site in Robinson R44 G-XZXZ. A bouncy castle was ballooning into shape and a climbing tower stood high, overlooking the controlled confusion.

The hangar interior was well organised and cleaned, with display boards and a diorama on that year's theme, the 70th anniversary of the Battle of the Atlantic. The Society's exhibits jogged memories, prompting visitors to share stories with the members on hand of aerial exploits and careers. Volunteers ran to and fro, parking cars, guiding tours, discussing displays and coordinating with the ever-helpful ATC cadets who once again joined in helping to marshal the event.

In the end, Ray's overnight

Open days were spectacular events, enjoyed by volunteers and visitors alike. Outside the hangar, the helicopter fly-in was a real treat.

CUT THROUGH HATCH
FOR EMERGENCY RESCUE

The Wildcat fighter, a restoration in progress, was a unique attraction at several outside events. The ability of the Society to bring exhibits on the road has proved to be a big boost to its public profile.

worries were unfounded: "Whenever it was over, and the last car went through those gates…the relief was just unbelievable."

The open day had been a smashing success, with over 4,000 visitors counted—a size which bordered on what could be safely and efficiently managed in a single day. The hangar and compound were barely cleared when the exhausted but overjoyed organising team

> The open day had been a smashing success, with over 4,000 visitors counted.

relaxed in the crew room with key volunteers and immediately figured that the following year's effort should be a two-day festival. But other meetings in other rooms elsewhere were entertaining selfish and limited political agendas. Their actions would ensure an open day would never happen in 2014 or for several years afterward. So what happened and why?

Simply put, Northern Ireland's two main political parties were

unable to agree on how the Maze/Long Kesh site should or should not be developed. The stalemate has continued to the present. It's not been made any simpler by the collapse of the Assembly. One direct consequence of the squabble has been an embargo on open days at the Society's hangars, with group tour parking limited to within the hangars compound.

As a non-sectarian, non-political charity, the Society

was both upset and confused by the turn of events, which left it effectively being held to ransom.

The only upside, if there was one, was the renewed media interest in the Society. The Northern Irish print and broadcast media, ever eager for a story with a political twist, jumped on the open days issue, unanimously on the side of the Ulster Aviation Society.

Year after year, Ray dutifully filled out applications for open days, but to no avail. Besides UAS discussions behind the scenes, he also spoke directly with the DUP's senior officials and with the deputy First Minister, Martin McGuinness of Sinn Fein. Again, the result was no open days. In fact, most applications did not receive so much as a polite answer from the OFMDFM. At the highest levels, the two partners refused to compromise; instead, each party blamed the other for the stalemate. It appeared that the Society's public outreach efforts were doomed. It was no small matter as well that the open days cancellations could also mean a cut in the Society's annual income of up to £15,000 a year.

The Society was financially sound for the moment, due largely to the success of the last open day and the slight increase in the organisation's presence at outside events, but the future looked rather shaky. As winter approached and the events and tours season settled, management committee paused to consider the year to come. The Wildcat had been a popular

The Buccaneer was one of several aircraft out in the hangar compound at the 2013 open day, but the only one with a performance routine. Those wings and airbrake were in action and the crowds loved it. Overall, it was a brilliant occasion…and the last one for years.

A BBC news crew showed up to report on preparations for the 2013 open day. The weather would be fine, but behind the scenes that summer some grey, political clouds were building.

draw at community festivals but its ongoing restoration meant that soon the aircraft would be too wide to cart around the country. The team needed something which could be easily transported and which would not be disruptive to the heritage projects.

As a non-sectarian, non-political charity, the Society was both upset and confused.

There might be an answer in a proposal put forward to the committee late in the summer: The Society should buy a Spitfire.

Well, not a real Spitfire. The price of such a legendary fighter was prohibitive— at least £1.5 million, plus

First Minister Peter Robinson and deputy First Minister Martin McGuinness enter a marquee at the Maze/Long Kesh site in 2013 to officially commission the body to run it. Both boasted of the site's great future. Within three months, the whole development idea was spiralling into collapse.

formidable maintenance, operating and associated annual costs. Anyway, it would be too valuable and cumbersome to take to community events. But why not a fibreglass replica? No restoration required, mobile and iconic, the aircraft would make an ideal events exhibit—a full-scale display aeroplane which could be assembled and dismantled for ground transport.

Some members balked at the idea of a replica—a dirty word almost to the more serious of aircraft collectors. But the majority supported it. After all, said one committee member,

A proposal was put forward in the summer: The Society should buy a Spitfire.

there were dozens of model aircraft in the collection; this was just an extra-big one.

A quick internet search revealed a cost varying from £50,000 to £80,000. That was still a bit rich for the Society's pocketbook. But as the committee mulled the acquisition, Treasurer Garry Macdonald spotted a second-hand example up for auction on the internet. Only one year old, it was located in Yorkshire and seemed a better replica than others the committee considered. They put in a tentative bid, and seller Neil McCarthy accepted

it. The terms were basic: £28,000 cash, delivered, and the money to be paid now.

Just as the committee was scrambling to figure out rapid payment, an anonymous Society member stepped up to offer an interest-free loan of £20,000, payable within five years. The £8,000 balance was easily available, so the committee snapped up the deal and voilá, they had themselves a big, toy Spitfire.

While waiting for delivery, Ray Burrows and Stephen Riley, the new public relations officer,

Twilight of a dream. The Society's hangars are behind the billboards, which spoke of hope long after it faded. The billboards are no longer there. The old prison is on the left.

popped over to the *Belfast Telegraph* to propose a marketing plan to editor Mike Gilson.

They painted a local history picture for him: The *Tele,* back during the Second World War, had organised a very successful public campaign to raise funds to buy a Spitfire. The resulting donations were high enough to purchase 17 of the fighters.

All were named after Northern Ireland communities, counties and regions, and under that inscription was the title: "Belfast Telegraph Spitfire Fund". But the only one actually based

in the province was a Spitfire IIa (P7823), named *Down.*

The Society could re-finish its new acquisition in the exact colour scheme and markings of P7823, and wherever it went the *Tele's* name and its remarkable campaign of years before would be seen by the public. Gilson was convinced. The company wouldn't provide a grant, but it would provide a massive amount of publicity by way of news and feature stories in the coming months, even years. (As it turned out, a flood of generous donations—private, corporate and through organisations—

Walter McManus, the Canadian pilot who died in the crash of the original Spitfire P7823.

enabled a payback of the loan within 10 months.)

The replica *Down* Spitfire, over time, proved to be a big hit wherever it went, nursed to success by a troop of dedicated volunteers who have cared for it, boosting the Society's profile at all sorts of events throughout Northern Ireland.

"Wherever we take that Spitfire, it just captures everybody's imagination," said John Martin, who has assisted the Spitfire team on many occasions. "People's eyes just light up... You're standing there beside a

The Society's Spitfire during its production in England. The aircraft was painted and marked in a livery quite different from its present appearance.

full-sized Spitfire that people can sit inside. It just transforms everything."

The aircraft arrived at the Society's hangar on December 22nd, 2013, a Christmas present whose graceful lines caught the eyes of everyone present. Somehow, despite the open days embargo, the coming year might not be that difficult after all.

It was time early that year, though, to mark the contribution of two major figures in aviation heritage. Society representatives took part in the unveiling of a monument to Harry Ferguson's

achievements, in the form of a beautiful metal sculpture of his magnificent monoplane, just off the northbound lane of the A1 at Hillsborough.

Sadly, the first weeks of the year also saw the Society members say goodbye to Sir Philip Foreman, who died peacefully on the 23rd of February. He was a patron to the group not just in title, but in action. "When I worked in Shorts, he was feared," remembered Ray Burrows. "He would have sacked you on the spot." However, when the committee got to know him in person, he turned out to be "a fantastic

The decision to buy a replica aircraft was radical but once it arrived there was no shortage of helping hands. It was just as well, for a brisk wind on arrival night threatened to fly the Spitfire up into its element.

Roy Kerr has a meeting of minds with a stubborn propeller. He won the struggle, and the props found their way back onto the SD3-30.

guy." Sir Philip's tireless support of the Society will never be forgotten; the lecture room on Hangar One which today bears his name is testament to that.

As springtime approached the restoration pace quickened, because soon the major efforts would turn toward preparing aircraft for increased hangar tours and participation in outside events. Shorts veteran Roy Kerr, saddened at the sight of the SD3-30 without its propellers, resolved to put that situation right. He laboured away out of sight at the north end of Hangar One, reviving the shine of the blades, no longer whirring in the sunlight but as brilliant as the day they were formed. Like all

volunteers, as Ernie Cromie never tired of saying, Roy would "carry on rewardless." Over at the big south-end doorway, a new arrival appeared that March in the shape of a Westland Scout helicopter, XV136. Though owned by the Air Training Corps, a deal was struck by which she would be maintained and displayed among the UAS collection.

Hangar One by now was becoming crowded, and covetous eyes were glancing next door toward Hangar Two, where the Society's vintage road vehicles were stored beside gym equipment, desks and foam mattresses from the derelict prison across the road.

The collection includes flight apparel dating back more than 70 years.

Those vehicles, like the aircraft, were targets for restoration. Mechanic Gary Millington started the engine of the old RAF's Bedford fuel bowser for the first time in 20 years. It was a noisy revival, but at least it was showing some heart. Gary, with occasional help from others, would spend three years at it before it looked like new with a loving lick of paint from Ian Hendry. Waiting nearby for Gary's future attention was another RAF airfield veteran, the huge Thornycroft Amazon salvage vehicle with its five-ton Coles crane.

Just keeping track of all the activity had become a serious chore. The management committee, hangar manager and

As the other teams fuss about with their aeroplanes, Gary Millington quietly restores the big RAF Amazon salvage machine to top condition. His start-up here looks impressive, if a bit messy.

the tours co-ordinator had to set priorities, order tools/parts/ materials, maintain the hangar, keep its landlord happy, liaise with editor and public relations, schedule trips and events and so on. All of these had to be synchronised, and many of them took place concurrently. It was all done by volunteers.

The major event for the spring of 2015 was weeks in preparation: The official unveiling and dedication of the replica Spitfire before it set about its work promoting the Society. TV presenter Wendy Austin, whose

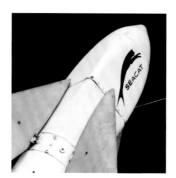

The Society collection includes a four-missile Shorts Seacat launcher.

father Cecil had been a wartime Spitfire pilot, had donated her father's logbook. The replica's livery had been replicated in the markings of *Down*, an aircraft of RAF 504 Squadron, based at Ballyhalbert, Co Down. The fighter had crashed on January 7th, 1942, north of Lurgan, killing its Canadian pilot, Walter McManus of the RCAF. Just over 100 members, guests and dignitaries attended the dedication event on April 25th, 2015, which came off very successfully. Second World War RAF veteran Fred Jennings proudly lifted the veil

to display the new markings, to the recorded roar of a Merlin engine and sustained applause from the audience.

The *Down* Spitfire was ready and the time had come to prove its worth. The Balmoral Show was to be the first in a packed events season which included the Lisburn Mayor's Parade, Belfast Maritime Festival, Moira Countryside Festival, Bangor Seashore Festival, Truckfest, the Newcastle Festival of Flight and, to cap the season off, the September Portrush air show. The replica delighted crowds wherever it went, its quality endorsed by the stunned reactions of onlookers who couldn't work out how the Society managed to transport by road an aircraft which appeared to be a real Spitfire. Of course, the transport was the easy part. The events team soon found that the assembly was not. The fighter could be dismantled easily enough in theory, but a crane was required to manoeuvre the heavy wings into place before and after each display. The events team was lucky to have on hand hauliers and members Matthew and Richard Belshaw whose lorry-mounted Hiab crane proved the ideal assembly tool.

By July, over 2000 members of the public had sat in the replica, shot their pictures and made a small donation to the Society. So impressed was one young lad at the Balmoral Show that upon exiting the cockpit he emptied the entire contents of his wallet into the donation bucket.

The Spitfire was an ongoing success—even a symbol of hope against the intransigence and childish behaviour of the politicians. As if to snort defiance in their direction, effervescent committee member Janette Kerr organised the first annual members' dinner at the Ramada Plaza, Shaw's Bridge, on August 30th. It, too, was successful.

Fulsome gestures of support might be seen through tributes to Paddy Malone and Ernie Cromie. Paddy received the UK Heritage Trophy for his development of the Royal

Observer Corps Rooms in Hangar One. Ernie, UAS chairman emeritus, received the prestigious British Empire Medal "for voluntary service to heritage in Northern Ireland through the Ulster Aviation Society". Ernie had kept himself busy since his retirement from the top UAS position, teaming up with Guy Warner to complete a trilogy of books on Northern Ireland aviation, the last of which—dealing with aircraft manufacture—was published in August, 2014.

The Society's members could be forgiven for not noticing, amidst the Spitfire triumph and the political pettiness of the open days embargo that a new aircraft had arrived on the hangar scene. How could they miss it, being the size it was? Puma XW222 had been donated to the Air Training Corps, but the cadets entrusted it to the Society for maintenance and management. By this time, the helicopter complement was fairly bloating the east side of Hangar One, much to the delight of visitors. They scrambled into the cavernous cargo holds of the Puma and Wessex, eight and ten at a time. They bulged into the Alouette, pressed their faces against the inside of the Perspex and whooped and laughed at the cameras and pretended like they'd never flown before.

The tour guides chuckled along with them, but among the restoration teams the helicopters' popularity faded a bit under the murmurs of a recurrent rumour that warmed the short days in the winter hangar. Another aircraft, it said,

Ernie's efforts to promote local heritage didn't go unnoticed. He was recognised with a BEM in 2014.

The Spitfire's specialist team has perfected the assembly task during the past four years, but this is how it began: A bit of measuring tape, a few tools and some puzzled looks. Joe Fairley, in the foreground, is wondering what welding will be needed.

The media were just as thrilled as the public with the Spitfire, which immediately began travelling by road to events throughout Northern Ireland.

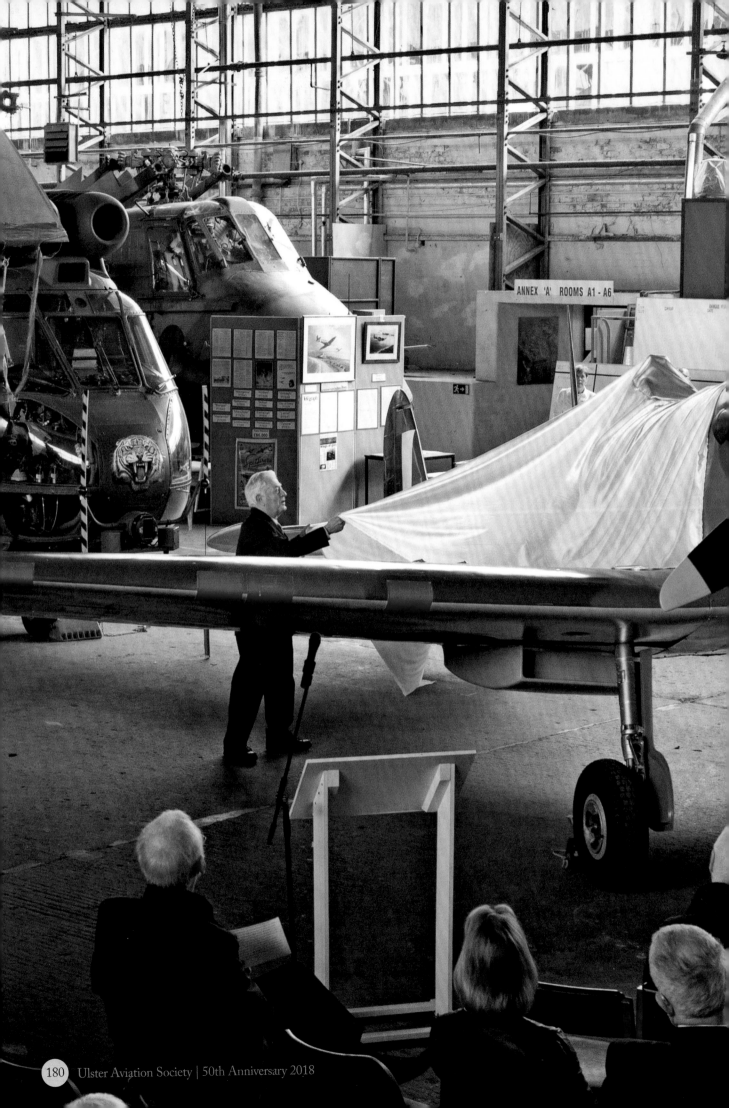

ROOF CLEARANCE 11'

HANGAR RULES

Health and Safety is everyone's responsibility
If you think something is not safe, make it safe
before you continue,
If you do not know how to ask someone who does!

LATEST NEWS

*Fred Jennings does the honours, unveiling
the replica Spitfire IIa in its new colours as
P7823, 'Down'.*

The massive Puma helicopter meets the small hangar door. That rotor head squeaked under the opening with easily three inches to spare. Or maybe just two. The Society is the custodian of the Puma, on behalf of the Air Training Corps.

"Hmmmm…What could we spend another £20,000 on?"

It just so happened that the committee had something in mind. The RAF base at Leuchars in Scotland was closing down and a couple of gate guardian Phantom FG.1s were looking for a new home. Groups of Society members, on their annual trips to the air show there, would have rushed past those Cold War legends many times, eager to catch the flying displays beyond.

Now, a heritage team made the journey once again and stopped near the entry gate, eyes fixed on XT864. Within weeks, their bid was made, the money paid, and the deal was done—thanks again to the generosity of Fred Jennings.

It would be a job on the scale of the Canberra and the team had made at least one decision up-front. No 'professionals' would be hired this time to take a big aircraft apart, working at a distance. A team from the hangars would do the job.

The Phantom would take a total of eleven working parties over the course of a year to plan and execute the collapse and transportation of the monster. Given that the Society was a group of volunteers, and that the Phantom wing joint was designed to be permanent, it's a wonder it didn't take more. But with the early guidance of Roy Kerr, a retired Phantom engineer, the team had the confidence to tackle the mammoth move. Several volunteers joined the effort at various times, among them Harry Munn, Leonard Craig, Jim Robinson, Mal Deeley, Geoff Muldrew, the Belshaws, David Jackson and Ray Burrows.

It took more than 1,000 man-hours of labour to dismantle the Phantom and return it to Northern Ireland. It was here, more than 40 years earlier, where XT864 and many of its stable-mates had returned from time to time for repairs and upgrading by the RAF's 23 Maintenance Unit (23MU). A hint of the Society team's exhaustive efforts at Leuchars

was waiting in the wings. It was somewhere out there, far away.

At about the same time, the anonymous donor of

the £20,000 Spitfire loan was "outed." Fred Jennings, unassuming and generous, would wander by the crew room occasionally, wondering aloud,

Phantom FG1 (XT864) in its RAF markings, standing ready at the entrance to the Leuchars base in 2015. Mats are in place to ease the task of towing to the hangar where dismantling can begin.

Three members of the Gannet restoration team persuade the front section to meet the middle section in 2017. Left to right are Anne McIlveen, Stephen Hegarty and Alan Moller. Other team members are doing their muscular bit elsewhere on the aircraft.

can be seen in an 18-minute video documentary, The *Phantom Phactor*, prepared by Stephen Riley and Mark Cairns. Still, 1,000-plus hours of effort reduced to 18 minutes? Merely a taste! And there was more work ahead. The big bird would be rebuilt and restored to its terrible beauty during the following months to prepare it for a showcase debut in the Society's 50th anniversary year.

Restoration work at the hangars was reaching a dynamic pace by this time—far beyond what any of the Society's emerging heritage faction would ever have believed back in the 1980s. Saturday was the busiest day, and it was not uncommon to see 50 volunteers or more

showing up to get stuck in. Besides the Phantom, the equally huge Fairey Gannet was consuming endless hours of work by its team. Certainly, the old maritime patrol aircraft needed it. The South Yorkshire Museum had donated XA460 to the Society in 2011, but it was—to be kind—in terrible shape. The tireless David Jackson, fresh from a pause to work just as hard on the Phantom, led a Gannet core team of about six, assisted by others on occasion, in a Hangar Two work area.

A new team had formed there as well around the wreck of a Fairchild Argus, picking up from where carpenter Gordon McCann

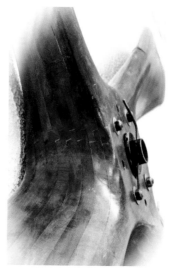

The art of restoration: This Handley Page propeller dates from a century ago.

had initiated work on the wing structures. He had left them in superb condition. However, much remained to be done before a completed frame could be delivered to specialist Owen Anderson for the fabric covering. Also sharing the Hangar Two space were a Canberra B2 nose, two Tucanos and the Thornycroft Amazon salvage lorry, each of them at various stages of restoration. By 2017, the buildings' landlords had recognised the Society's need (and expanding occupancy) of the second hangar, so included it in the licence. It was one less headache to deal with, for both parties. The Wildcat, enjoying determined but sporadic attention from Jim Robinson and Harry Munn, was promptly

David Jackson is lead hand on the Gannet crew. A retired engineer, his expertise has been an asset to the Society on many occasions—including turns with two other 'heavies', the Canberra and the Phantom.

Paul Trimble oversees the crane work as the Argus fuselage leaves its nest in a barn near Moira in 2012. G-AJSN ended its flying career in 1967 in civilian colours, snapping a wing in a ground loop at Cork Airport. It's nearing completion, restored by the Society as an ATA aircraft in WW2.

William Smyth and son Nathaniel in 2017 work on the wing ribs of the Society's Sherpa, part of a long, meticulous process to bring this one-of-a-kind aircraft up to display condition.

rolled in from Hangar One. The move provided more room there for displays, leaving the Sherpa as the only major restoration in that building. Its team was smaller, but the work required an especially fastidious touch—not least because its distinctive and delicate, wooden wings were being built totally from scratch, without the missing original drawings to depend upon.

It can't be stressed enough that the restoration work was a fundamental feature of interest to tour groups. Not only could visitors get an appreciation of the demands confronting the work teams, they could ask questions and discuss progress with the people directly involved in the efforts. It was first-hand immersion in history—not the kind of opportunity they would get at an active aerospace production line.

In fact, the two hangars were themselves works in progress. In recent years, there were major repairs to the windows, electrical system and ventilation. There were new toilet facilities (and the facilities before that? Don't ask!).

An aircraft engine display section for Denis Piggot's amazing machines was fashioned from one of the dingy and cluttered side rooms, and Ernie Cromie put the finishing touches to the self-explanatory Aldergrove Room. Other facilities were turned into a boardroom, an administration office—with a couple of desks! —clean and bright storage rooms, and a First World War display room. Fred Jennings'

radio room was renamed in his honour, and his library moved to larger quarters, which gradually shrunk as hundreds of donated books arrived—to make a total of over 5,000 by 2018. A larger crew room with almost-new seating was set up, but increased volunteering had already made it too small. That too was a symptom of expansion.

Visitors, especially those on return trips, were beginning to notice. School tour groups were increasing in numbers and size, several repeating their bookings year after year. Education of the general public is the aim of any museum facility, but targeting by schools is especially welcome. By 2016, class groups numbered in the dozens. Schools with STEM (Science, Technology, Engineering, Mathematics) curricula were especially interested.

The expansion wasn't limited to volunteer numbers. The visitor intake was becoming progressively higher. A decision to personalise each tour with a guide had meant a related toll in 2015 upon the existing pool of volunteers. It prompted repeated calls for more hangar guides to cope with the pressure. Thankfully, the call was answered and Ernie quickly trained his growing team of guides in time for another busy visiting season. That training became increasingly more comprehensive in recent years. Clearly, the guides were required to verse themselves in the basics of the Society, the individual aircraft and displays. There was training in health and safety (as with all volunteers), child protection protocols, how to work with guests of various ages or disabilities. There was always at least one volunteer in place with training in first aid.

A couple of educational groups needed no guides. One was the Air Training Corps' Northern Ireland Wing, to which the Society had given space in Hangar Two in 2017 for their own project. The young folks, under professional guidance and a grant from Boeing, were assembling an aeroplane. Specifically, it was a TL-Ultralight Sting, a sporty two-seater made of composite materials. They were aiming to have it ready for flight in 2018, the centenary of the Royal Air Force. Nearby, a room adjoining the hangar was being prepared for the arrival of regular classes from Belfast Metropolitan College. The occupants would be diploma students in aeronautical engineering, featuring hands-on experience with aircraft from the collection. And this, it would do well to remember, was in a facility which some politicians wished to see punished as a form of protest. Compounding that embargo was the Society's licence with its various restrictions, including the nuisance factor of a six-month renewable term. Official documentation makes it clear that the six-month limitation was originally imposed due to government uncertainty about development of the total Maze/Long Kesh site. Thirteen years later, the uncertainty continues and so does the six-month provision. The Society was also forbidden to advertise.

Guided tours remain a staple of Society activity, engaging all kinds of social groups, schools, and professional organisations and involving thousands of visitors each year. Ron Bishop has guided many of those groups; he is seen here beside the Society's V-1 replica flying bomb.

The Society encourages visitors to climb aboard several of the exhibits, in this case the Alouette helicopter donated by the Irish Air Corps. These students, on a 2015 tour, were from St. Bride's Primary School in Belfast, eager to pack as many as they could onto the flight deck.

Skyward from Magilligan

Local heritage loves a strong symbol.

There is no better example of Northern Ireland's remarkable aviation heritage than the Society's Ferguson monoplane replica. Built here in 2015-2016, it linked the present with the dawn of powered flight just over a century earlier.

Harry Ferguson—he of farming tractor fame—was the first person in Ireland to design, build and fly his own powered aeroplane. That achievement took place at Hillsborough in December, 1909.

But back then powered flight was largely an eccentric fancy, and most of the pioneers' original aeroplanes disappeared long ago.

So the question arose: Why not imitate history?

The British Broadcasting Corporation approached the Society in 2015, looking for some hangar space for the assembly of a replica. The idea was simple: The BBC would commission the building of the aeroplane. Concurrent filming for a three-part television documentary would relate the progress from construction right through to—hopefully—its maiden flight. Dick Strawbridge, a popular television presenter originally from Northern Ireland, would follow the process from beginning to end. At the same time, he would tell the parallel story of Harry Ferguson's aviation exploits. William McMinn, a local pilot with a few hundred hours on his logbook and adventure on his mind, offered his services not just to fly the machine, but to put his engineering experience to use in guiding its assembly. Local firms would build the various components, as close as possible to one of Harry's original designs. As concessions to safety, modern instruments and a reliable, up-to-date engine would be installed.

William picked Society member Stephen Lowry to help him. During that winter and into 2016, the pair nursed components into a configuration that eventually began to look like the frame of a wood-and-wire wonder. And all of this under the prying lights and prodding lenses of BBC cameras.

The pair worked well as a team, said William: "Stephen had as much commitment as I did. We pushed each other all the way."
Stephen himself was thrilled to be part of the experience.
"There'll be memories for myself here," he said. "And I can say to my grandchildren: 'I built that.'"

For its maiden flight, the sky-bride would need clothing. The man to provide the fabric covering was specialist Owen Anderson up at his shop by the Ulster Gliding Club's field at Bellarena, Magilligan Point. This was the very location where Harry Ferguson himself had test-flown versions of his machine over a century before.

Owen patiently stitched and glued the cloth tightly over the frame and had the Fergie Flyer—as some observers were calling it—ready by early May. It received regulatory approval and, after a few weather delays, headed into a slight breeze and took to the air on May 15th, 2016.

It was powered flight, certainly, but it wasn't sustained and not easily controlled. No turns, no banking, a quick climb to only 20 feet, then a hard but safe landing. And it was brief—maybe 10 seconds' duration. But, under William's skilful handling, it worked! The aeroplane flew, and more than that: It was, in a sense, a flight back to the past. The whole experience allowed the teams of designers, the builders, the ground crew, the BBC and above all the pilot, to dare the challenge and to savour a brief but keen sense of what the aviation pioneers of more than a century ago must have experienced.

The flight was too tenuous to suggest that the fragile replica might ever fly again. The teams that designed, built and introduced it to the sky may never again know that unique, first-flight shiver of joy. But the memory of that amazing adventure belongs to them for life.
And the replica? It belongs to the Ulster Aviation Society.

(The television documentary, The Great Flying Challenge, *was broadcast on BBC Northern Ireland in three weekly segments in September and October, 2016)*

Harry Ferguson, aviation pioneer.

Stephen Lowry, left, assisted pilot William McMinn through the demanding assembly and flight preparations of the Ferguson replica. Both are Society members.

Peter Johnston (right), director of BBC Northern Ireland, with Society chairman Ray Burrows. The date is September 28th, 2016, when the BBC donated the Ferguson aircraft to the collection.

We've got liftoff! With the hills of Donegal in the background, William McMinn guides the replica through its brief, historic flight from the field at Bellarena, Magilligan.

Roy Kerr (standing, left) oversees a workshop on health and safety. The hangars are fully stocked with health and safety equipment in strategic locations and many of the volunteers are trained at different levels of first aid.

It begs the question of how a museum facility could attract the public if people weren't aware it was even there. It has been a point raised continually by surprised visitors, many of them resident within a mile or less of the hangars. The Society has approached the problem with a basic understanding: It does not pay for advertising. On the other hand, the news media have been quite aware of the Society's collection. Over the years, television crews have visited the hangars for features and news stories, educational inserts, drama and children's

> It begs the question of how a museum facility could attract the public if people weren't aware it was even there.

programming and so on. News reporters in print media in particular have been tireless and supportive in regular coverage of the continuing stalemate over the open days issue. Such reports have done the government and politicians no favours, while at the same time keeping the Society's positive activities and reputation in the spotlight. On the other hand, as some of the organisation's leaders have noted, increased public awareness has meant increased hangar tours, which have put further pressure on the guides' workload. That has

resulted in regular recruitment campaigns. It's the kind of dilemma many museums would love to deal with.

The hangars became popular venues for a variety of groups not necessarily associated with education or aviation. An international club of DeLorean owners convening in Belfast brought a cavalcade of the unusual cars to the hangar compound and spent most of an afternoon there. The Cedar Foundation, a charity for the disabled, paid regular visits, then endorsed the Society for

The Society's relationship with the Air Training Corps is an excellent one. The cadets were given space to build their Boeing-sponsored Sting microlight, completed early in 2018. Interested cadets are also encouraged to join the Society.

a major national award. In an annual evening event, the Belfast Film Festival hosted over 200 for films with an aviation theme (*Con Air, Top Gun* and *Airplane!* for example). Even later in the evening on several nights through the year, health and safety officer Mal

Deeley hosted a paranormal group which connected with the spirit world. One hangar volunteer swore he saw an aircraft glowing for a moment, perhaps as a tribute. It was the Phantom. No surprise there. Another group, one which forged a special connection with the

It's the kind of dilemma many museums would love to deal with.

Society, was Northern Ireland's Polish community. Ernie Cromie has regularly attended their annual remembrance event on November 11th at Milltown Cemetery in Belfast, where several airmen were buried during the Second World War. As a matter

Another example of fulfilling the Society's education brief: engineering students from Belfast Metropolitan College learn some basics of aircraft jacking from their teacher, Alex Philp, who happens to be an active Society volunteer.

A television crew at work on a feature production for Irish-language channel TG4 from the Irish Republic, with a school group touring the hangar in 2017.

of special importance, the date also commemorates Poland's Independence Day. A social occasion, featuring a documentary on the Polish contribution during the war, has been held in the hangar and other events have been held in Belfast in recent years, with the *Down* Spitfire and Society members in attendance. (Two Polish fighter squadrons were actually based for brief periods at RAF Ballyhalbert in Co Down during the war, including No. 303, the highest-scoring fighter squadron during the Battle of Britain).

Another Polish connection featured in 2015, as the spirited Janette Kerr organised a members' excursion to Krakow, home of the Polish Aviation Museum and its 200-plus aircraft. It was an overdue resumption of trips, and 24 members and friends took advantage of it. Another committee member, John May, took note of the positive reviews from members and resolved to continue such outings.

Closer to home, the events team lined up another impressive schedule that year, boosted this time by invitations to Kilkeel's GI Jive and the Dalriada festival in Glenarm. And, ever ambitious, the team also decided to try out the Alouette helicopter at the Balmoral show. They discovered to their delight that it was every bit as popular as the all-star Spitfire. With no on-site assembly needed, the Irish Air Corps helicopter was easily transported and proved an excellent companion to the legendary fighter.

The pair teamed up again for the big show of late summer: Air Waves Portrush.

The biggest external display the Society had ever staged required no fewer than four show-pieces to be transported all the way to the north coast: Spitfire, Alouette, Wildcat and Canberra B2 nose. The Society's contribution was such that the festival's organisers invited it to become a partner for the future years. It was a satisfying reward for a huge effort by the events team. Sadly, a "home event" never happened at all: The 2015 application for an open day at the Maze/Long Kesh hangar brought the same response from Stormont: No answer at all. And therefore, no open day. Maybe the following summer. Still, the Society's participation at outside events and tours made 2015 a successful year, and a key activity crowned the volunteers' efforts: It was time to bring Phantom XT864 home. After a huge effort to separate the wings, the fuselage was shipped to Hangar Two on June 12th. Five months later, the bulky wing edged under the shutter door as well. The Phantom foray to Scotland and back was complete.

"No civilian organisation, never mind a voluntary charity, has ever achieved what we have," said Ray Burrows. "We all should be justifiably proud."

If 2015 had been the year of expansion (more visits, more events, more trips, more aircraft) then 2016 was the year of evolution. Partly through reputation, partly

Otto, the iconic blow-up doll from the movie Airplane! *bids goodnight to the Belfast Film Festival audience. Hangar Two has been their annual venue in recent years. It's another example of the diverse in-house activity which the Society has been encouraging.*

though their own initiative, the Society diversified into as wide a range of activities as it had ever undertaken.

The 2016 events season kicked off early with a new addition to the calendar. However, the Shane's Castle Steam Rally suffered from poor weather, resulting in a disappointing weekend out for the events team. Still, weather in the north being what it is, the events team was philosophical about it. There were other outside compensations. One was a trip for the replica Spitfire to the Oval, Glentoran Football Club's home ground, to commemorate the Belfast blitz. A bomb had blown out the centre of the pitch in 1941. The Spitfire also starred in a jaunt to the Guildhall

The learning never stops. One of hundreds of technical and instruction manuals in the Society collection.

Square in Derry/Londonderry to mark the anniversary of the U-boats' surrender in 1945.

However, just up the coast, potential success turned to serious disappointment at the Air Waves Portrush event. Despite its new status as a partner in the festival, the Society's allotted site was behind a huge marquee which hid the displays and the aircraft exhibits from potential visitors. Ray Burrows had a strong discussion with event organisers at the conclusion of the event, promising no return by the Society unless the situation was corrected for 2017.

A late outing for 2016 came in November when Richmond Nursing Home in Holywood

Back to the Past: It's 2016, and more than 60 DeLorean automobile owners visited the collection. It was clear and unusual evidence that Ulster's engineering heritage (and the sheer joy of collecting) extends well beyond aircraft.

Spring and summer have become very active seasons for the Society, with a roadshow featuring aircraft and displays to as many as 20 community events. Paul Young, here at Kilkeel's GI Jive Festival, fills in a few details about the legendary Spitfire and its amazing history.

asked if the Spitfire could attend a family day there. It was an unusual request, but one the team were happy to fulfil: Richmond had been home to Sir Philip and Lady Margaret Foreman in their last days.

Since the death of Sir Philip, the Society had been without a patron. The committee had toyed with replacements but had not pursued the matter until one day, RAF Air Commodore Harvey Smyth popped into the hangar. Here was a possibility: a respecte local (Banbridge) fellow and Society supporter. He was also

> ## Since the death of Sir Philip, the Society had been without a patron.

a friend of Chairman Burrows, who recalled the appointment ceremony—such as it was.

"Just as he was walking towards the door, I said, 'Excuse me, Harv, we were wondering: Would you consider becoming the Society patron?'" He would be honoured, he replied. And Ray said he, too, was honoured—and so was the Society. All appointments should be so simple.

On September 16th, 2016 the new patron dropped in unannounced, his first visit in

that capacity. It happened to be the wedding day of secretary Stephen Hegarty, who followed the air commodore into the hangar with his new bride, Kirsty, to have their wedding photos taken. The newlyweds and the new patron were introduced, a speech or two was delivered, and all parties left with smiles on their faces.

The hangar crew were still smiling as well, as carpenter extraordinaire Peter Duff put the final touches on a small but outstanding restoration project: the repair and re-

Aviation enthusiasm inspires performance! Songwriter Fionnuala Fagan has penned several poems to music, motivated by stories of Ulster's aviation legends past and present. She has performed in Hangar One for Society volunteers.

finishing of a rare propeller from a Handley Page V/1500. The precise laminations and shine of the century-old piece of hardware were something to behold. Once completed in Hangar Two, the four-bladed wonder was moved back from the shutter door to make way for another newcomer: a large van, beautifully covered in Society livery. The committee had decided that, in lieu of another major aircraft acquisition, a van was needed to increase the independence of the events team. Up until then, they'd been relying on contracted hauliers for every

event they attended. Though still dependent to some degree on larger lorries, the van would ease the physical and financial burdens somewhat.

A major test for the van presented itself in October. The earlier committee decision to forego any aircraft acquisitions in 2017 fell flat on its nose when an aeroplane came looking for them. Aeroventure, the South Yorkshire Aviation Museum which had provided the Gannet, had another surplus machine to give up. This was a Percival Sea Prince, WF122, which had actually

Carpenter extraordinaire Peter Duff put the final touches on a small but outstanding restoration.

served at Sydenham as a Royal Navy navigation trainer.

So, a definite Northern Ireland connection. Who could resist? A crew of five—Harry Munn, Jim Robinson, Tom Lennon, Steve Lowry and Ray Burrows—piled into the new van and headed for England on October 17th. They returned six days later with the big, twin-engined brute only hours behind on two lorries. They had also managed a side trip to Manston, and snug aboard the lorries as well were two massive wing tanks and an ejection seat for the Phantom.

Banbridge man Harvey Smyth had a few questions about the Collection's Gannet, in this chat with restorer Billy McCall (left). The RAF Air Commodore (now Vice-Marshal) was intrigued with the complex wing-fold system of the naval ASW aircraft.

The treasure trove was quickly unloaded and hauled into Hangar Two.

Next door in Hangar One, the Society by now was routinely handling school groups of up to 60 students. The guides' training and personal touch were paying off, with some schools booking year after year. By the autumn of 2017, tour numbers for all visiting groups had hit a new high with 145 bookings totalling 3,900 people, not counting the informal 'walk-ins' that crossed the Society's threshold.

The positive comments in the visitor book at the reception area, combined with the

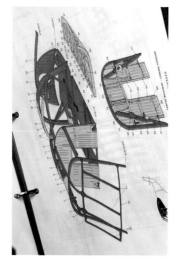

Detailed drawings from the Society's technical library have proven helpful to the restoration teams.

number of tours, produced a significant coup for the Society. Thanks largely to the guides' efforts and hospitality, the Society's collection inched ahead of all other attractions in the Lisburn area to become number one in TripAdvisor's popular list of go-to locations. And as this book goes to press it's still at the top of that list.

The 50th anniversary of the Ulster Aviation Society was looming. The management committee formed an ad hoc task force to prepare for 2018. Its planning had begun in the spring of 2017. The group comprised Mark Cairns, Alan Chowney, Stephen Hegarty,

Garry MacDonald, Des Regan and Stephen Riley, with John Weir volunteering at a later stage.

The first project of the task force was a calendar for 2018.

Charles McQuillan, an award-winning national feature photographer, offered his services to provide the 12 images, each one featuring an aircraft from the collection. Charles had built a relationship with the Society while shooting the Ferguson aircraft assembly and flight project. Through his lens, the ordinary had become art, and with Mark Cairns' able design work the calendar was a sellout through three printing runs. The 50th subcommittee was clearly focused on the future but the hangar crews still had much work facing them in 2017.

In Hangar Two, ace painter Ian Hendry had finished his work—for the moment—on the Wildcat that spring. Ray Burrows had hoped that Peter Lock, who had ditched the aircraft in Portmore Lough in 1944, might make it to the hangar for another visit from his home in Canada. Peter had not been well, but Ray knew it would boost his spirits to see his old bird in its brilliant Fleet Air Arm livery—or at least the port side of it, as the starboard wing was still being restored.

The visit was not to be. The Society learned in May of 2017 that Peter had died in January. He is fondly remembered by the friends he made here in Northern Ireland.

Spotters at heart: Three of our stalwart travellers get an eyeful of a see-through Mirage F1 fighter at the Le Bourget Airport museum in France. Left to right are Bert McGowan, Harry Munn and Alan Chowney.

attend, the major one being the Royal International Air Tattoo in England. The RAF had sent the invitation, asking that the Spitfire team bring the Society's replica to grace the main banquet hall. It was a quick case of there and back, for the Air Waves Portrush event was on again in a matter of days. The persistent John Weir reminded everyone that he'd been pressing to get the huge Puma helicopter up to the north coast festival, among all the other goodies. Yes, said the management committee, and yes again to the Wessex, both the big whirlybirds bound for the biggest annual event of Northern Ireland's aviation year. It was also the largest display—six aircraft, a huge marquee with a popular shop and brand-new graphics—that the Society had ever mounted for an outside event. The weather was a bit grim for the flypasts, with cancellation of some star acts, but earthward the

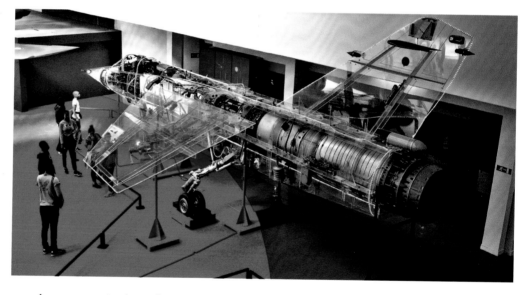

crowds were ecstatic about the ground exhibits. The Wessex and Puma took top honours, with more visitors than the other aircraft combined.

A few days later came a hint of future hope. The elected politicians at Stormont, frozen solid on a simply Maze/Long Kesh agreement, had even been unable to sustain a working government. That left the civil

> There was restoration to continue and events to attend.

service to do what it could to maintain basic services and make a few decisions. One of those judgements was a last-minute approval for limited Society participation in European Heritage Open Days. It was actually only one day, September 8th, with hardly any public notice and with extremely restricted parking. No fly-ins, no bouncy slides, no bands, and yet the visitors came. Only a

Phantom XT864 has become the collection's Cold War flagship. Society Patron Harvey Smyth and former Phantom pilot Chris Bolton unveiled the famous fighter in its Royal Navy colours on April 28th, 2018.

few hundred could make it, but volunteers showed up by the dozen. The effort was deemed a success as far as it went, and no problems were evident.

Those factors alone might bode well for a full return to open days, said the optimists. And, in the end, that 50-year-old spirit of optimism still thrives in the Society.

Even while the last guests were filtering out of the compound gates, plans were well in hand for the Society's big knees-up in the anniversary year. Founders Night was planned for September 29th as a salute to all the volunteers in the organisation's history. Live music, a tasty menu, terrific prizes and great craic would mark the evening. It featured a royal garnish: The official presentation by the Lord Lieutenant of County Down of the Queen's Award for Voluntary Service. The citation gives the Ulster Aviation Society credit for "preserving and exhibiting Northern Ireland's aviation heritage with skill, dedication and passion."

There was a solemnity about the occasion, however, at a moment that joined past and present in a poignant companionship of bereavement. Founding chairman Eddie Franklin, who welded an informal little family of spotters into a tight band of enthusiasts 50 years ago, passed away on August 12th, 2012.

And only days before the Founders Night, Harry Munn,

a smiling, extraordinary, hangar volunteer and dear friend to all who knew him, died suddenly at his home in Lisburn.

A 50th anniversary is a fitting place to take pause and be thankful that for half a century there has been a community of friends like Eddie and Harry and so many others, giving freely of their time to grow this Society into a home where everyone is welcome to work, learn, share and if nothing else, have fun. At the same time, of course, there's a future waiting.

> "Looking to the future I am convinced that our next 50 years are secure, and will be as glorious as the last."
> – Air Vice-Marshal Harvey Smyth

The Society's site at Air Waves Portrush in September, 2018. The large, camouflaged Wessex and Puma helicopters were especially popular for the climb-in crowd. Just out of sight behind the marquee are the Ferguson flyer and the replica Spitfire.

The wrecks are still out there. This tail section of a Meteor NF.14, once based at Aldergrove, lies stuck in a hedge near Glenavy.

Argus in the autumn, with colours to match the season, courtesy of painter Ian Hendry. When completed, the aircraft will be represented as an Air Transport Auxiliary machine, flown by at least two female pilots during the Second World War.

Fifty years is worth celebrating in itself. But winning the Queen's Award for Voluntary Service made 2018 extra special. Here (left to right) members Charlie Taggart, Jack Woods, Ernie Cromie, Ray Burrows and David Hill join County Down Lord Lieutenant David Lindsay (centre) on receipt of the award. The cadet is Colour Sgt Toni Walker, RIR, aide to the Lord Lieutenant.

The Queen's Award
for Voluntary Service

2018

5.1 | A Hangar Tour

There's a knack to opening the door of Hangar One. The awkward handle has baffled many first-time visitors. But the reward is that heart-stopping moment when the door swings away and you're left staring into a cavernous space.

One aircraft after another fills the void. It's early morning and the light from the rising sun floods through the high east windows, striking each machine, casting shadows and odd shapes the whole length of the enormous building. It's quite the first impression, and one which still takes the breath of veteran Society volunteers to this day.

But beyond the vista of Hangar One, many hidden treasures await the curious mind. So join us now on a journey behind the aircraft into the less-documented corners of the Ulster Aviation Society collection.

We stroll towards the hangar reception desk and our diligent membership secretary, Des Regan. Rarely seen without his orange high-visibility vest, no visitor may pass into this hallowed space without a word of greeting from Des. A quick introduction, a short health and safety briefing and we're on our way. We pass the crew room, a portable cabin adapted as the volunteers' rest area, fondly referred to as 'the crew room'. Inside, the workers joke and laugh over a cup of tea. We'll not disturb their moment's calm. Passing by, we turn behind the imposing Wessex helicopter and reach our first stop: the Ernie Cromie Room. Named for the past chairman of 30 years, it's an ode to Ernie's passion, the Second World War in Northern Ireland.

Display boards around the walls tell the story of the USAAF during the war while the glass cases in the centre house some of the collection's most precious exhibits. In one is the control panel from a Short Stirling. Very few relics of these mighty bombers still survive. Nearly 2400 of these aircraft were built, most of them in Northern

Ireland, yet not a single, complete example remains. In the room's centre stands a "drop tank", thousands of which were fabricated from thick, pressed paper by Northern Ireland printers Nicholson and Bass. Slung from the wings or belly of fighter aircraft, such fuel tanks greatly extended their range during the war. One display case is dedicated to the hobby of scale plastic models, while ration books and aircrew training manuals adorn another. As we move towards the back of the room we find tributes to local pilots who served during the war, like Coleraine's Victor Beamish, group captain and Battle of Britain ace. Displayed are his medals, among them the Distinguished Service Order and bar, and the Distinguished Flying Cross. He was listed as missing in action on March 28th, 1942.

Through an adjoining door, we enter the First World War Room which is well worth a visit. Carrying on from the theme of the Ernie Cromie Room, local pilots are remembered for their bravery and leadership. Their stories are told on displays around the room's walls.

At the room's centre stands a glass-topped table in which is told the tale of the Tyrrell brothers of Belfast, RAF pilots aged 19 and 23.

One of few surviving relics from the colossal V/1500 bombers built by Harland and Wolff, this propeller is a reminder that Ulster's aviation heritage is almost as old as powered flight itself. It was restored beautifully by Society craftsman Peter Duff.

Visitors to the Ernie Cromie Room explore the local impact of the Second World War. Thousands of Nicholson and Bass paper fuel tanks like the one in the foreground were built in Northern Ireland, symbols of the many local contributions to the war effort.

The First World War One Room commemorates local airmen who never came home from the 'war to end all wars.'

They were killed in action within eleven days of each other in June, 1918.

Dominating one wall are two massive, wooden propellers of the First World War era. One is from a BE.2c, the other from a Belfast-built Handley Page V/1500. The latter has been beautifully restored by volunteer Peter Duff.

An adjacent hallway leads to three rooms dedicated to the Royal Observer Corps. Curated by ROC veteran Paddy Malone, these rooms recreate a ROC headquarters and a diminutive ROC bunker in which three-man teams would monitor equipment for the first signs of a nuclear strike. Items like the ground zero indicator provide humbling evidence of just how close the world came to oblivion during the Cold War.

Leaving the hallway, we venture back into the nave of Hangar One, towards the north gable. On the way we pass the Sir Philip Foreman Room, an educational and lecture space named for the Society's first patron. The room, complete with classroom chairs and desks, donated by Queen's University, also incorporates a full set of audio-visual aids, thus serving as a theatre space as well.

Carrying on down the hangar we notice that the whole building's immense steel frame rests upon concrete pillars four feet high. Ostensibly a standard T2 wartime hangar, the Long Kesh examples had to be raised on these stanchions in order for the huge Stirlings that were assembled here to pass through the hangar doors and underneath an overhead gantry.

Before we reach the far gable, we'll duck into another little hallway to our right. Here we find the Society library and the Fred Jennings Radio Room. Fred is a highly-respected volunteer and war veteran who, despite reaching his ninety-third year, continues to curate the radio display and manage the library when he's not guiding a tour or out with the events team.

The library is home to more than 5000 books, donated and collected over 50 years, on all subjects aeronautical. There are publications on the pioneers of flight and the design of spacecraft. There are collated volumes of such magazines as *Flight International* and *The Aeroplane* and dozens of more obscure publications. A recent addition is *The Complete Book of the SR71*, signed by author Col. Rich Graham who came to address the Society here in Long Kesh in May, 2018.

Across the hall in the radio room we'll find aircraft communications equipment from the Second World War to the present day. Displays explain how aircraft navigated in wartime, complete with original copies of navigators' astronomical tables and observation instruments.

It also contains the Society's own amateur broadcasting station. Managed by David Gregg, the Society transmits on special occasions under the handle GB4UAS. Opposite the radio station lies the highlight of the space, an active digital radar display showing real-time aircraft flight paths over Northern Ireland.

We'll leave Fred's domain now and cross the hangar floor to the eastern wall. On our way we'll pause by Mal Deeley's display of survival equipment. The focus of this collection is a row of ejector seats from Martin Baker, founded by Ulsterman James Martin. Beside them stands a mannequin dressed in a modern g-suit with flying helmet and life vest, just one outfit in the Society's collection of distinctive flying clothing. Completing the line-up is a cabinet

Paddy Malone, dedicated curator of the Royal Observer Corps Rooms, stands beside a direction-finding blast gauge. He's holding a blast power indicator. Original exhibits like these are solemn reminders of how close the world came to the brink during the Cold War.

Society members listen attentively as Ernie Cromie introduces speaker Air Vice-Marshal David Niven. The Sir Philip Foreman Room serves equally well as a classroom for visiting schools and a cinema for the Society's video productions.

The development of aircraft systems has been every bit as dramatic as the advances in aerodynamic and engine technology. In the Fred Jennings Radio Room visitors can study aircraft radios of all eras, and even watch incoming aircraft on the room's local radar display.

Aircraft/Vehicles

1. Blackburn S.2B Buccaneer
2. Supermarine MkIIA Spitfire Replica
3. Hawker FB.5 Sea Hawk
4. de Havilland T.11 Vampire
5. BAC T3A Jet Provost
6. Embraer/Shorts Tucano
7. English Electric/Shorts PR.9 Canberra
8. Quicksilver Ultralight
9. Aerosport Scamp
10. Shorts SD-330
11. Shorts SB.4 Sherpa
12. Ferguson TEA 20 Tractor
13. Ferguson MkII 1911 Flyer Replica
14. Evans VP-2
15. Rotec Rally 2B Microlight
16. Himax R-1700
17. Clutton-Tabenor Fred Series 2
18. Fieseler V-1 Replica
19. Westland AH.1 Scout
20. Aérospatiale SA 316B Alouette III
21. Bedford Ql Fuel Bowser
22. Robinson R-22
23. Air&Space 18A Gyroplane
24. Aérospatiale HC1 Puma
25. Westland Wessex HC.2
26. Percival P.57 Sea Prince T.1
27. Fairey ECM.6 Gannet
28. McDonnell Douglas FG.1 Phantom
29. Grumman F4F Wildcat
30. Fairchild 24W-41A Argus
31. Shorts Tucano - Fuselage
32. de Havilland C2 Devon - Nose Section
33. English Electric B2 Canberra - Nose Section
34. 1936 Dennis Big Four Fire Engine
35. Thornycroft Amazon Crane
36. Pitts S-1A Special
37. Piper PA-32-300 Cherokee Six
38. Aero Gare Sea Hawker
39. Taylorcraft Auster
40. Henri Mignet Flying Flea
41. Shorts Tucano Static Test Article
42. Eurowing Goldwing

A snapshot of our collection as it stands in late 2018. But as the Society evolves, so aircraft and items join the collection, are rearranged within the hangars or are loaned to events and exhibitions nationwide.

THE UAS HER

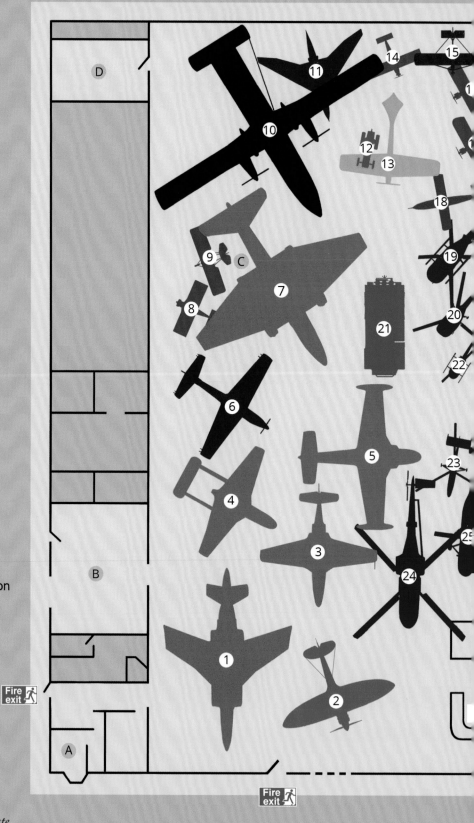

👫	Toilets	🍴	Food
♿	Disabled Toilet	🔥 Fire exit	Fire Exit
👶	Baby Changing	➕ First aid	First Aid
ℹ	Information	🔋	Defibrillator

E COLLECTION

Rooms and Displays

A VIP Room
B Engine Room
C Drones and Missiles
D Aldergrove Room
E The Fred Jennings Radio Room
F Library
G The Sir Philip Foreman Room
H Royal Observer Corps 1
I Royal Observer Corps 2
J Royal Observer Corps 3
K First World War Room
L The Ernie Cromie B.E.M. Room

Future Projects Not on Display

Fire exit

E
F
G
I
K
L

27
26
28
29
30
31
32
33
34
35

36
37
38
39
40
41
42

Legend

- Fast Jets
- WWII
- Helicopters
- Prop Driven
- Shorts Bombardier
- Vintage
- Home Builds
- Cockpits
- Support Vehicles
- Miscellaneous

of items included in aircraft survival packs, from flares and stoves to rations and knives. Modern aircrew must be prepared for all eventualities.

From here we'll move towards the Canberra's tail, behind the SD3-30 to the hangar's latest display, the Aldergrove Room. Another of Ernie's contributions, the room tells the story of Aldergrove from the First World War to the present day. The location was used at various times as a manufacturing site, RAF Coastal Command station, army base and of course, Belfast International Airport. All eras are explained and the display is completed with a large architectural model of the airport as it could have developed but for an economic downturn. Another highlight of the room is easy to miss. A wall-mounted A4 certificate appoints one William Henry Armstrong to the role of pilot officer in October of 1930. Mundane, perhaps, but for the original signature of His Majesty George V in the upper left-hand corner.

Leaving the Aldergrove Room, we proceed along the west wall of Hangar One. On the way we'll pass large and small unused rooms. Some are stores for future exhibits, others are vacant, waiting for the day they'll be transformed into museum-class exhibition spaces. But another room on this side is a treasure trove for the wandering engineer. Denis Piggot is in charge here, running the Engine Room with help from Joe Fairley and John Dunphy. It's home to the majority of the Society's aircraft engine collection.

In-lines and radials, pure jets and turbo-props, all styles and eras of engines are represented here. There's the legendary Merlin, recovered by the Society from Cushendall in 1984. There's the monstrous 14-cylinder Pratt and Whitney Twin Wasp radial, the type which powered our own Wildcat along with many other designs, among them the ubiquitous Douglas DC-3 (C-47, Dakota). We'll see the Astazou-XII turboprop, troublesome powerplant of choice for early variants of the Shorts Skyvan.

Display cabinets in the room contain other engine paraphernalia like RPM and fuel gauges, but one cabinet is special. Here we'll find the only items in the Society' collection from Concorde, the Mach 2 marvel. They're only small, but the turbine and compressor blade sections are enough to cause the visitor to stop for a moment and consider the technical achievement behind this engineering wonder. This space is also a workshop as Denis and the team restore their powerplants. Perhaps in the future this hangar will again hear the sound of running aircraft engines. For now, however, the team carries

on quietly, every day adding more and more for the visitor to see.

The very history of aviation is scattered throughout Hangar One and its annex rooms in the shape of more than 100 scale model aircraft, the products of several of our talented members. Several of these precision pieces are competition winners, crafted and finished to represent in the most authentic manner all sorts of flying machines.

Behind the bulk of the Shorts-built Canberra PR.9, in the northwest corner of Hangar One, is the home of other products from the

Aldergrove has been a hub of civil, military and aviation manufacturing for generations. The collection's Aldergrove Room explores the past, present and future of Northern Ireland's largest airport.

ENGINE ROOM

The entrance to a land of motor magic: The Society's Engine Room is at once a museum site, workshop and restoration area. It's an exciting stop for visitors on their hangar tours. Pistons, jets and turboprops are all represented, and if a member of the engine team is in residence, he'll be happy to explain the wizardry that makes them move.

company, including the Shorts SD3-30 (G-BDBS). Behind it is the SB4 Sherpa, undergoing restoration; a model of the aircraft and a cutaway section of its wing indicate clearly what made the Sherpa a unique design. There's a line-up nearby of large demonstration models from Shorts. Some represent aircraft from their product line, others are examples of what might have been.

There's a hallway running past the ROC Rooms on the east side of Hangar One which joins it to Hangar Two. This hangar is chiefly a restoration space, but impossible to miss in the centre of the building is Phantom XT864, fully completed. Behind it, graphics boards tell the story of the RAF's 23 Maintenance Unit at Aldergrove where almost every British Phantom was serviced and maintained. Next to that sits an engine from the dawn of the jet age. The Rolls Royce Derwent from a 1940s Meteor is paired with the 1960s Rolls Royce Spey of the British Phantoms, making a striking demonstration of the pace of engine development during the Cold War. And of course, the Wildcat restoration area is completed with a display of artefacts recovered from the wreck. Nearby graphics displays portray and explain the whole recovery operation.

Other major aircraft projects currently in hand in Hangar Two include the Gannet, the Argus and the Sea Prince. Hangar Two is always in a state of flux, with new arrivals coming while aircraft under restoration are relocated from time to time to meet the changing needs of working and display space. The only annex room in use is the Belfast Metropolitan College's workshop on the east side of the building.

Some areas of Hangar Two are cordoned off for safety purposes, but most of the restoration work is clearly visible. Visitors can wander about outside the barriers and chat with members of the restoration teams; they're always happy to answer questions. Or you could return to Hangar One and walk among the aircraft themselves, learning about how each one links to some part of Northern Ireland's long aeronautical history. Maybe you'd prefer to revisit some of the display rooms and take a little more time to explore the depth of the Society's interpretive exhibits. Throughout this whistle-stop tour, one thing should be absolutely clear by now: The Society's enthusiastic volunteers are proud of Ulster's aviation heritage.

The Society's large model collection lets visitors explore aircraft otherwise absent from the collection, as well as the liveries of various local and international operators. Above is a Short 3-60.

Look at these guys: What a bunch of winners! It's easy to think of heritage as a long distant touchstone, but here former fitters of Alder-grove's 23 Maintenance Unit reunite for an event at the hangars of the Ulster Aviation Society. The collection provides an outlet for people to share their stories and show that history is real and tangible.

The actual aircraft types make the first impression to the wide-eyed visitor, but it's the models, artefacts, displays and photographs that give them context and truly connect the machines to their Ulster home.

CHAPTER SIX

Nobody at the birth of the Ulster Aviation Society could have planned the half-century evolution which has resulted in today's dynamic organisation. They never saw that coming. Just as certainly, of course, most of the next half century is beyond the members' planning capabilities.

Still, dreaming has brought us this far. It will surely be a companion in the future. Just as surely, among the members of the UAS will be those who work to transform dreams and ideas into realities.

In the near future, we hope to see one or two aircraft—maybe even more—arriving at the hangars, demanding attention from the restoration teams. As if they weren't busy enough!

Still with the hangars, the popularity of the collection has increased noticeably in the past two or three years, and that seems bound to continue. Hangar Boss (and Chairman) Ray Burrows continues to plead for more Society members to step forward as volunteer tour guides. This will ease the burden on the current cadre of guides, as well as introducing new faces to a fascinating adventure.

It seems likely that UAS membership will reach 600 in 2019, and each person will continue to receive *Ulster Airmail* as part of membership privileges. The magazine has undergone remarkable changes

The spotters can still be spotted: What began as informal meetings of UAS members in 1997 has evolved into a photography-focused club, of special interest to spotters. The Belfast Aviation Group is independent of the UAS, but includes several members of the Society.

> It seems likely that UAS membership will reach 600 in 2019.

in two years, so a bit of a breather may be in order! However, change never sits still, so editor Graham Mehaffy is always on the lookout for ideas—especially for content. Graham is an avid modeller as well, and the work of that select group will hopefully become even more important as hangar displays change and new ones are added.

In fact, all the exhibit rooms are bound to be improved and/or complemented as the years go by. New items will be added and new ideas will come to the fore. It's just the nature of change.

As chairman, Ray is determined that the changes will be well managed.

Eddie Franklin left the UAS as heightened security restricted access to his beloved Aldergrove. But he never fell out of love with aviation. A confirmed bird watcher, he was unable to resist a quick snap of this sculpture marking the birth of powered flight in the UK while on a "birding" excursion to the Isle of Sheppey.

"I know some people are beginning to get worried that we are getting so popular we won't be able to control it," he said recently. "I don't want to get to the stage where our success is running away and we can't keep the reins on it."

That might mean, in light of a continued increase in guided tours, that hangar visits will be limited to certain days. In the larger scheme, outside the museum site itself, it might mean a limit as well to the number of outside events which the UAS attends. But it will likely mean more variety in the aircraft and displays, new faces in the volunteer ranks and maybe even some breathing time for the Spitfire's elite assembly crew.

"It would be nice to have an 'A' team and a 'B' team and a 'C' team for the Spitfire," Ray muses. Spreading the load, that would mean only five events or so for each aircraft team in the touring season. Quite civilised, actually.

He's certain that, for the foreseeable future, there will be work to keep the aircraft restoration teams busy. Even as new aircraft arrivals threaten to fill both hangars to capacity, the "old timer" machines—such as the Sea Hawk—will be requiring some serious overhaul and maintenance work.

And while that's going on,

> Ray is certain that, for the foreseeable future, there will be work to keep the aircraft restoration teams busy.

Natural Lighting

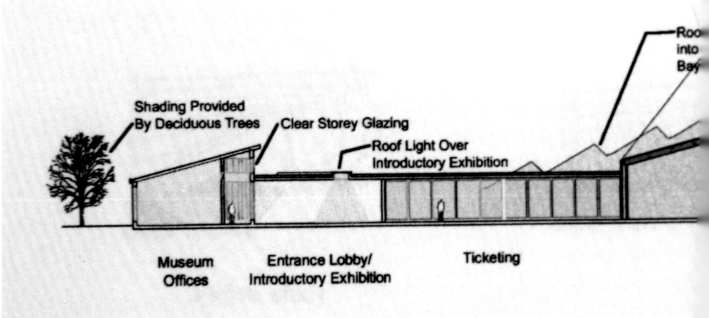

Roo
into
Bay

Shading Provided
By Deciduous Trees

Clear Storey Glazing

Roof Light Over
Introductory Exhibition

Museum
Offices

Entrance Lobby/
Introductory Exhibition

Ticketing

The Ulster Aviation Society has been a source of inspiration well beyond its own membership. Queen's architecture student William A.J. Dawson imagined a bright, modern home in 2004 for Ireland's largest aircraft collection.

there's a growing challenge to improve the interactive aspects of the hangar, with a graphics strategy and other possibilities.

Among them might be a hands-on air traffic control setup, a virtual flight deck, an imaginative setting for photo shoots in the hot-air balloon basket and so on.

"All sorts of things like that," says Ray. "But we need somebody to actually process ideas."

> "That is our secret. We are engaging with the public. They're coming in their droves to visit us."

Those are all challenges for the future's short term, but they could run on for years. For the moment, excitement is beginning to build for an ambitious trip proposal in the summer of 2019. John May is planning a tour to run a week (or maybe more?) to several aviation museums and perhaps an air show or two in the northwest United States and a pop across the border to Canada as well. That would be the most impressive bit of travel in the Society's history.

It's an exciting component in what will be another busy year for the Society's management committee. But during 50 years of success amidst a welter of problems, the committee's example of leadership has proven an inspiration and shows no signs of abating.

As for the long term, what will the next half-century offer? Maybe, if government bureaucracy and politicians get considerably more positive, a return to open days. Maybe

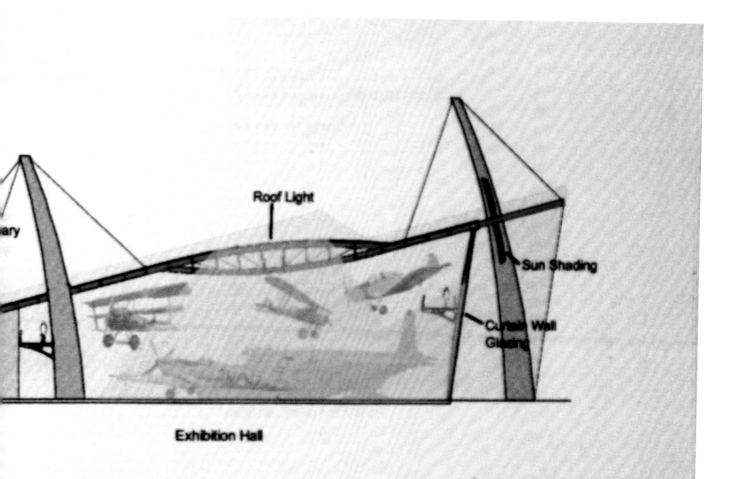

Roof Light

Sun Shading

Curtain Wall Glazing

Exhibition Hall

full museum accreditation to the hangars. That's a change Ray and many others would welcome, as a move to instil tight standards of best practice. Perhaps there could even be a change in the hangars' legal limitations. The hangars are licensed—and have been since after the move from Langford Lodge—only as mere storage facilities, with a stingy, six-month renewal term. That's not the kind of thing to impress big granting outfits like the Heritage Lottery Fund when the Society goes looking for sizeable financing.

Ray has had to restrain his language when searching for the words to express his annoyance at trying to amend the licence or deal with inflexible government, which controls the Society's licence and the land of the larger Maze/Long Kesh site.

"I don't know when they'll get the message," he sighs. "We have something special here. The ordinary, everyday person gets the message. Why don't they get the message: Wake up. Wake up!"

It's a fleeting yet ever-present frustration. For solace, he and the other Society activists recognise and indulge with regularity in the enthusiasm and hard work which have been the foundation of the Society's half-century of success. And the reward is enlarged by dealing with the public.

"To me, that is our secret," says Ray. "We are engaging with the public now, through taking aircraft out, through our website and Facebook. They're coming in their droves to visit us. That is our secret. We need to continue with engagement."

Fifty years from now, that continued spirit and the enthusiasm of volunteers within the Ulster Aviation Society may guarantee another half-century of enjoyment and success.

It's one thing to imagine the future, but quite another to lay the groundwork. Besides aircraft restoration, members are constantly painting and refurbishing the hangars' many annex rooms as new displays or functions evolve. This one ended up as a workshop.

Society members have often been invited to RAF Aldergrove to see visiting aircraft—Tornadoes among them. Perhaps soon, the UAS hangars will be the destination of tourists and enthusiasts hoping to see this fast jet at close quarters.

50 Years Ago:
Technology Rewind

Just as dreaming and energy have been companions throughout Society history, another one has been technology.

Some perspective is in order. Fifty years ago, there were no footsteps on the moon. Television, still in relative infancy, would soon show us those feeble lunar impressions. There was no Youtube to do that job on personal computers, which didn't exist anyway. There were no smartphones, electric cars, instant cashpoints or robotic vacuum cleaners. Tweets came only from birds.

There was no digital photography for spotters to share worldwide on the Society's Facebook page within seconds of clicking a camera shutter.

Trevor Franklin, son of founder Eddie, has recalled the unwieldy process back in the 1970s which his father used to produce aircraft prints from negatives.

"He turned the bathroom into a darkroom to develop the pictures. He'd take the film out of the camera in a bag, and he made like it was all a trick," he said. "I'd help him sometimes and there was something magic about the developing, how he did it and make the prints on an enlarger. It was all magical when I was young."

With maturity and experience, we're not so innocent now of the potential of technology, for better or for worse. Still, it's a very complicated phenomenon, and too often left by the rest of us to the wisdom of those few who seem to understand its best uses. To the Ulster Aviation Society, it will present many opportunities to present our product in new ways to attract new audiences. An in-house flight simulator? Touchscreen interpretive displays? Perhaps the air traffic control desks of Sydenham and Belfast International, rescued from the scrapheap by Ray Burrows, will be brought back to life in an interactive challenge for visitors young and old. Whatever form it takes, it will certainly be fun—perhaps even a momentary return to the innocent kind of magic which founder Eddie Franklin performed for his young son.

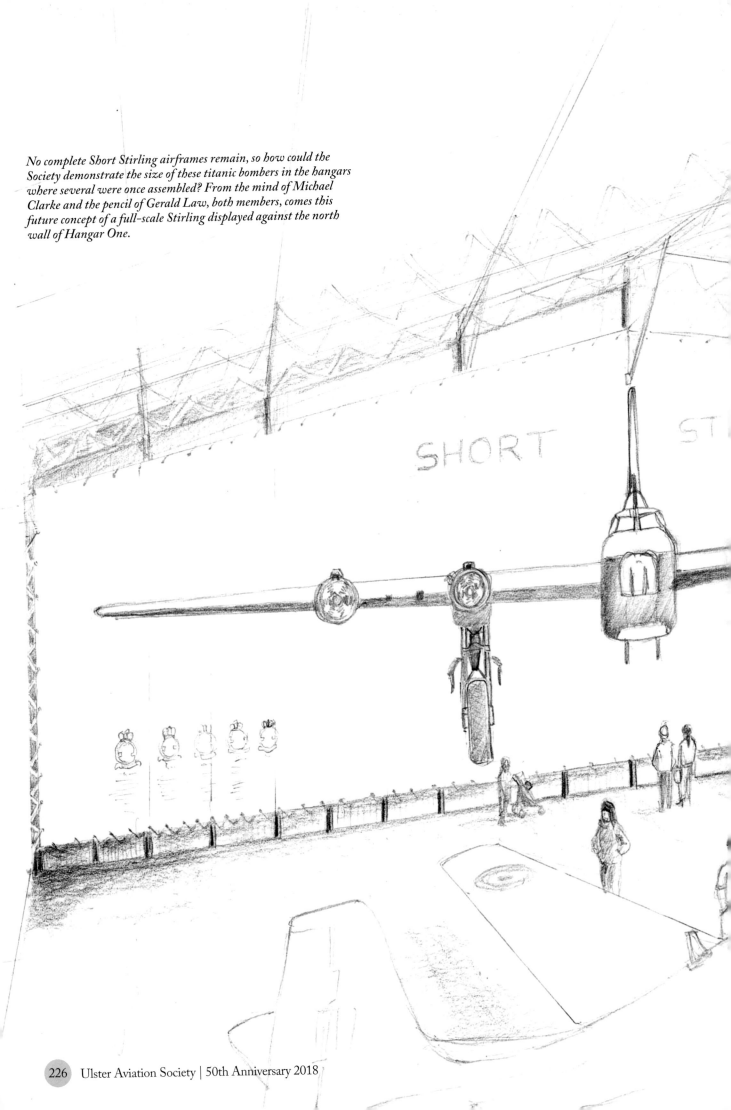

No complete Short Stirling airframes remain, so how could the Society demonstrate the size of these titanic bombers in the hangars where several were once assembled? From the mind of Michael Clarke and the pencil of Gerald Law, both members, comes this future concept of a full-scale Stirling displayed against the north wall of Hangar One.

Appendix 1: The Ulster Aviation Collection

The contents of the Ulster Aviation Society's hangars are too many to list. In fact, there is an ongoing battle to catalogue each item in the ever growing collection. But among the books, uniforms, models and medals, there are the 40 aircraft which form the Society's 'shop front.' Each has a local link, some stronger than others, and each has its own story to tell.

#	Manufacturer	Model	Serial	Note	Status	Local Connection
1	AeroGare	Sea Hawker	EI-BUO		Work in progress	Owned by Irish pilots Charlie Lavery and Chris Donaldson
2	Aérospatiale	HC1 Puma	XW222		Displayed	Served in Northern Ireland with 230 Sqn. Owner: N.I. Wing, ATC.
3	Aérospatiale	SA 316B Alouette III	202		Displayed	Operated with the Irish Air Corps
4	Aerosport	Scamp	Unregistered		Displayed	Owned by local Don Chisholm (deceased)
5	Air & Space	18A Gyroplane	EI-CNG		Displayed	Owned by Irish pilot Pat Joyce
6	BAC	T3A Jet Provost	XM414		Displayed	Owned by Lisburn and Castlereagh City Council
7	Beagle	A.61 Terrier	G-AVCS		Future project	Bought by member John May following a ground loop at Cranfield
8	Blackburn	S.2B Buccaneer	XV361		Displayed	Buccaneers were serviced at RAF Sydenham
9	Chargus	Cyclone	BAPC.263	Hang glider	In storage	Owned by local Billy Reed
10	Clutton-Tabenor	Fred Series 2	G-BNZR		Displayed	Owned by local Mervyn Waugh
11	De Havilland	C2 Devon	VP957	Nose section	Work in progress	Nose section saved and refurbished by the Northern Ireland Wing, Air Training Corps
12	De Havilland	T.11 Vampire	WZ549		Displayed	Vampires flew with 502 (Ulster) Sqn. RAF and Irish Air Corps and were often seen in Sydenham
13	Embraer / Shorts	Tucano	G-BTUC	Prototype	Displayed	Modified by Shorts
14	English Electric	B2 Canberra	WF911	Nose section	Displayed	Built under licence by Shorts
15	English Electric / Shorts	PR.9 Canberra	XH131		Displayed	Built under licence by Shorts
16	Eurowing	Goldwing	G-MJWS		Future project	Owned by local Jeff Salter (and first flew from Langford Lodge)
17	Evans	VP-2	G-BEHX		Displayed	Owned by local George Adams
18	Fairchild	24W-41A Argus	G-AJSN		Work in progress	Operated and ground-looped in Cork.

19	Fairey	ECM.6 Gannet	XA460		Work in progress	Gannets served with 719/737 Sqns at HMS Gannet (Eglinton)
20	Ferguson	MKII 1911 Flyer	G-CJEN	Replica	Displayed	Replica of Harry Ferguson's monoplane
21	Fieseler	V-1		Replica	Displayed	Built by member James Herron
22	Grumman	F4F Wildcat	JV482		Work in progress	Based at RAF Long Kesh with 882 Sqn
23	Hawker	FB.5 Sea Hawk	WN108		Displayed	Served as instructional airframe at Shorts' training school
24	Henri Mignet	Flying Flea		Fuselage only	Future project	Northern Irish built example
25	Himax	R-1700	G-MZHM		Displayed	Built and flown by local Maurice McKeown
26	McDonnell Douglas	FG.1 Phantom	XT864		Work in progress	Phantoms were serviced at RAF Aldergrove's 23 Maintenance Unit
27	Percival	P.57 Sea Prince T.1	WF122		Work in progress	Operated with Sydenham's Station Flight
28	Piper	PA-32-300 Cherokee Six	G-CDUX		Future project	Crashed at Newtownards
29	Pitts	S-1A Special	N80BA		Future project	Owned by locals Tom Stronge and Patrick Gallagher until a non-fatal crash in 1999
30	Quicksilver	Ultralight			Displayed	Replicated Harry Ferguson's Newcastle flight with member Ernie Patterson at the controls
31	Robinson	R-22	G-RENT		Displayed	Owned by local Harold Hassard
32	Rotec	Rally 2B Microlight	G-MBJV		Future project	Owned by local Stuart Wilson
33	Shorts	SB.4 Sherpa	G-14.1		Work in progress	Unique research aircraft built by Shorts
34	Shorts	SD-330	G-BDBS		Displayed	Shorts production prototype
35	Shorts	Tucano	ZF167		Work in progress	Built by Shorts
36	Shorts	Tucano		Test article	Future project	Built by Shorts
37	Supermarine	MkIIA Spitfire	P7823	Replica	Displayed	Original purchased with money raised through the *Belfast Telegraph* Spitfire Fund
38	Westland	AH.1 Scout	XV136		Displayed	Served in Northern Ireland during Operation Banner with the Army Air Corps (NI Wing ATC)
39	Westland	HC2 Wessex	XR517		Displayed	Served with 72 Sqn in Northern Ireland during Operation Banner
40		Rogallo Glider	BAPC.266	Hang glider	In storage	Owned by Charles Linford, former secretary of Ulster Hang Gliding Club

Appendix 2: Flightpath

1968 On **December 15th**, a small group of Northern Ireland based aircraft spotters broke from the Irish Society of Aviation Enthusiasts and, under the leadership of Eddie Franklin, founded the Ulster Aviation Society.

1969

February
Ulster Airmail #1 published, edited by Hugh McGrattan

June 10th
Pink Floyd plays the Ulster Hall

July 20th
Apollo 11 lands on the moon

August 15th
400,000 music lovers attend Woodstock

1970

February
The UAS begins to sell Ian Allen aircraft markings books

March 21st
Dana wins the Eurovision Song Contest

April 17th
Apollo 13 returns safely to earth

May 26th
The Tu-144 becomes the first commercial Mach 2 aircraft

1971

November
Woodgate Aviation begins to print *Ulster Airmail*

The Benny Hill Show launches on ITV

April 19th
Salyut, the first space station, launches

October 1st
Disneyland resort opens in Orlando

1972

January
Roger Andrews takes over as editor of *Ulster Airmail*

The Floral Hall in Belfast closes

January 30th
Trouble in Derry/ Londonderry sparks 'Bloody Sunday'

March 30th
Northern Ireland parliament suspended

1973

May
UAS members visit the site of downed Corsair JT693

May 14th
Britain abolishes capital punishment

October 19th
Libya embargoes oil exports, launching the fuel crisis

1974

September
First annual UAS trip to the Leuchars air show

Invention of the Rubik's Cube

April 9th
Richard Nixon resigns following the 'Watergate' scandal

August 22nd
First flight of the Shorts SD3-30

1975

July
The UAS purchase Woodgate Aviation's Gestetner duplicator

April 30th
The Vietnam War ends

June 20th
Jaws opens to huge audiences

July 31st
The Miami showband are ambushed

1976

December
Heightened airport security impacts the UAS spotters. Meetings are halted

January 21st
First commercial flight of Concorde

July
Shorts Sandringham *Southern Cross* visits Belfast

1977

October 18th
UAS meetings restart at Antrim Technical Collage

September 10th
Con Law flies three members to the Leuchars air show

Home computers hit the popular market

May 25th
Star Wars Episode IV hits the silver screen

August 16th
Elvis Presley dies

1978

June
Due to the lack of *Airmail* articles, the UAS is close to collapse

Gloria Hunniford starts on *Good Evening Ulster*

February 14th
HMS *Ark Royal* is decommissioned

October 16th
Karol Wojtyla becomes Pope John Paul II

230

Timeline

1979
- **September 16th** — UAS meetings move to Ards Flying Club
- **July 1st** — The Sony Walkman debuts
- **May 4th** — Margret Thatcher is elected Prime Minister
- **October 27th** — Mother Teresa wins the Nobel Peace Prize

1980
- **May 18th** — Mount St. Helens erupts
- **May 22nd** — Release of *Pac-Man*, the video game.
- **December 8th** — John Lennon is murdered

1981
- **September 26th** — UAS attend their first Ulster air show
- **April** — First appearance of the UAS logo
- **April 12th** — The space shuttle Columbia is launched

1982
- **January 12th** — Ernie Cromie is elected Society Chairman
- **April 2nd** — Argentina invades the Falkland Islands
- **December 23rd** — First flight of the Shorts C-23 Sherpa

1983
- **December 3rd** — First of several Society discos
- **October 5th** — Wildcat JV482 engine recovered
- **February** — Members fly with the Air Atlantique DC-3
- **July 18th** — Sally Ride becomes the first American woman in space

1984
- **April 30th** — Wildcat JV482 recovered from Portmore Lough
- **January 24th** — Steve Jobs reveals the first Macintosh computer
- **October 26th** — Bishop Desmond Tutu receives the Nobel Peace Prize

1985
- **December 15th** — UAS members fly in Trident G-AWZU
- **January 7th** — Wildcat pilot Peter Lock visits the UAS
- **November 20th** — Microsoft introduces Windows

1986
- **November** — Shorts SD3-30 G-BSBH donated
- **October** — The UAS sell out five Air Atlantique DC-3 flights
- **January 28th** — Space shuttle Challenger explodes shortly after lift-off
- **April 26th** — The Chernobyl nuclear meltdown occurs

1987
- **November 8th** — Aviation artist Iain Wylie addresses the members
- **July** — The Society and Newtownards Council begin to plan a museum of flight
- **March 30th** — A Van Gogh "Sunflowers" sells for £27.75 million.
- **May 28th** — West German pilot Mathias Rust lands a Cessna 172 in Red Square

1988
- **December 27th** — The UAS acquire Vampire WZ549 from RAF Coningsby
- **November 8th** — George Bush Sr. is elected president of the United States
- **December 21st** — Pan Am Flight 103 is bombed over Lockerbie, Scotland

1989

- **October 14th** — Sea Hawk collected from Shorts' training school
- **September 22nd** — Members travel to the Leuchars air show
- **June 4th** — Protesters killed in Tiananmen Square
- **November 9th** — The Berlin Wall falls

1990

- **February 11th** — Nelson Mandela is freed from prison
- **October 4th** — Bombardier acquires Shorts
- **August 2nd** — The Gulf War begins

1991

- **October** — The UAS collection moves to Langford Lodge
- **June 22nd** — Members fly from Dublin to Shannon by Boeing 747
- **August 6th** — The Internet becomes available for unrestricted use
- **December 25th** — The Soviet Union is dissolved

1992

- **October** — Monthly meetings move to CIYMS
- **June** — *Ulster Aviation Handbook* published
- **January 25th** — UAS members fly on a Saab 340B
- **January** — Plans for an Ards flight museum fall through
- Red Bull, an "energy drink," goes international.
- **February 7th** — The European Union is created
- **November 3rd** — Bill Clinton is elected President of the USA

1993

- **April 7th** — SD3-30 G-BDBS airlifted by Chinook to Langford Lodge
- **December 2nd** — The Hubble telescope is repaired in space

1994

- **April 5th** — Buccaneer XV361 flown to Langford Lodge
- **April** — Debut of *Ulster Airmail* in A5 format
- **April** — Genocide and civil war take place in Rwanda
- **May 6th** — The Channel Tunnel is opened

1995

- **May 20th** — First UAS open day, attended by B-17 *Sally B*
- Pokémon is created in Japan
- **February 21st** — Steve Fossett finishes the first trans-Pacific balloon solo
- **June 29th** — Space Shuttle Atlantis docks with the Mir

1996

- **September 29th** — A UAS team film of the last Vickers Merchantman flight
- **March 26th** — Red Ten, of the Red Arrows, addresses the UAS members
- **August 26th** — Princess Diana and Prince Charles get divorced

1997

- **November 18th** — Society members vote to become a charity
- **March** — The UAS website is established
- **March 22nd** — The Hale-Bopp comet makes its closest approach to Earth
- **September 25th** — The search engine Google is founded
- **December 19th** — *Titanic*, the worlds costliest movie, released

1998

- **April** — *In the Heart of the City* book launch
- **April 10th** — The Good Friday Agreement is signed

1999

May 2nd
Langford Lodge heli-meet

February 12th
President Clinton is acquitted of perjury and obstruction of justice

August 17th
An earthquake in Turkey kills over 17,000

2000

September 24th
Langford Lodge heli-meet

September 2nd
Members fly with Aer Lingus DH84 Dragon *Iolar*

March
Restoration of Langford hangars ongoing

February 11th
The Northern Ireland assembly is suspended

July 25th
Air France Concorde F-BTSC crashes after take-off from Paris

October 31st
First astronauts arrive on the international space station

2001

September 8th
Langford Lodge Vintage Extravaganza

June
Colour introduced to *Ulster Airmail*

January
Tucano G-BTUC joins the UAS collection

February 12th
Draft of the human genome published

September 11th
Terrorist attacks on the World Trade Centre and the Pentagon

October 26th
The Lockheed Martin X-35 wins the Joint Strike Fighter contest

2002

June 16th
Langford Lodge Wings and Wheels day

January 1st
The Euro is introduced

July 3rd
Steve Fossett completes the first solo round-the-world balloon flight

November 13th
The MV *Prestige* oil tanker spills its cargo off the coast of Spain

2003

December 20th
Jet Provost XM414 arrives at Langford Lodge

August
Sir Philip Foreman becomes the UAS Patron

February 28th
Robinson 22 G-RENT loaned to the Society

February 1st
The Space Shuttle Columbia disaster

March 20th
The second Gulf War begins

May 30th
The final commercial flight of Concorde

2004

August 14th
John Miller concert at Aldergrove

April 19th
UAS receives the notice of lease expiration

March 30th
Wessex XR517/N arrives at Langford Lodge

February 4th
Facebook launched by Mark Zuckerberg

December 21st
Over £26m is stolen from a bank in Belfast city centre

2005

Restoration of the Maze/Long Kesh hangars begins

January 21st
The last items are moved to the Maze/Long Kesh

Jan-Dec
Members begin moving exhibits out of Langford Lodge

April 27th
First flight of the Airbus A380

July 7th
London bombings on underground trains and a bus kill 56

August 29th
Hurricane Katrina hits New Orleans

2006

August 24th
Pluto is expelled from the planetary club

December 30th
Saddam Hussein is executed

2007

October 23rd
Eric 'Winkle' Brown addresses the UAS

September 13th
UAS take delivery of a Himax 1700R

August
Northern Ireland's Battle of the Skies DVD released

Crocs, the colourful plastic shoes, are the year's fashion statement

February
Start of the global economic crisis

June 29th
Apple releases the iPhone

2008

September 27th
First Maze/Long Kesh members' open day

September 1st – 3rd
Avro Vulcan talks mark the UAS 40th anniversary

July 17th
Shorts SB.4 Sherpa arrives at Long Kesh

October 21st
The Large Hadron Collider is inaugurated

November 4th
The US elects its first black president, Barack Obama

2009
- **August 18th** The launch of a series of events marking the 100th anniversary of Harry Ferguson's flight
- **May 29th** Irish Air Corps Alouette 202 donated to the Society
- **March 11th** Shorts-built Canberra PR.9, XH131, purchased
- **September 30th** Earthquakes in Indonesia kill at least 1000
- **October** The oldest human ancestor is found in Ethiopia

2010
- **December 13th** The Canberra arrives at the Maze/Long Kesh
- **January 12th** Haiti is struck by a devastating earthquake
- **April 20th** The *Deepwater Horizon* platform explodes causing an environmental crisis

2011
- **December 14th** Fairey Gannet XA460 parts arrive at the UAS hangars
- **June 11th** Open day at the Maze/Long Kesh
- **March 11th** Japan is devastated by an earthquake and tsunami
- **July 21st** The US space shuttle fleet is retired

2012
- **August 25th** Society holds USAAF themed open day
- **March 3rd** Fairchild Argus G-AJSN joins the UAS collection
- **March 27th** Ernie Cromie steps down as Chairman, succeeded by Ray Burrows
- **February 6th** The Diamond Jubilee of Queen Elizabeth II
- **July 27th** Opening Ceremony of the London Olympic Games
- **December 3rd** Violent protests in Belfast over the frequency of flying the Union Flag

2013
- **December 21st** A replica Spitfire is purchased
- **August 24th** Battle of the Atlantic themed open day at the Maze/Long Kesh
- **May 15th** The Wildcat goes to the Balmoral Show, restarting UAS public event appearances
- **February 15th** A meteor explodes over Chelyabinsk, Russia
- **May** The first creation of human embryonic stem cells by cloning
- **September 16th** First flight of the Bombardier CSeries, with Belfast built wings

2014
- **November** UAS successfully bid for Phantom XT864
- **September** Open days cancelled: government permission is not received
- **March 2nd** Westland Scout XV136 arrives at the hangars
- **March 8th** Malaysia Airlines Flight 370 disappears en route to Beijing
- **September 18th** Scotland votes "no" to independence
- **November 12th** The Rosetta probe deploys its lander 'Philae' on comet 67P

2015
- **November 22nd** The Phantom arrives at the Maze/Long Kesh
- **January 7th** Terrorist attack on Paris' *Charlie Hebdo* magazine offices
- **August 22nd** A Hawker Hunter crashes at the Shoreham air show
- **September 10th** Queen Elizabeth II is the longest reigning British monarch

2016
- **September 1st** Belfast Film Festival show 'Con Air' in the UAS hangars
- **August** Wing Commander Harvey Smyth becomes the UAS patron
- **May 15th** The Ferguson Flyer replica flies at the Ulster Gliding Club
- **April 8th** SpaceX successfully land a Falcon 9 rocket booster
- **June 23rd** The UK votes to leave the EU
- **November 8th** Donald Trump is elected president of the United States

2017
- **October 13th** Sea Prince arrives at the Maze/Long Kesh
- **May 20th** Members travel to Speyer, Germany to visit local museums
- **January 6th** Wildcat pilot Peter Lock dies
- **January 10th** The Northern Irish Government collapses
- **July 14th** Grenfell Tower in London catches fire, killing 71
- **October 5th** Worldwide support for the #metoo movement against sexual harassment

2018
- **September 29th** Founder's Night: a celebration of 50 years and the Queen's Award for Voluntary Service
- **May 9th** Col. Rich Graham (rtd), SR71 pilot, addresses the UAS
- **April 28th** Restored Phantom XT846 unveiled in Fleet Air Arm livery
- **March 25th** First commercial flight direct from London to Perth
- **April 27th** Historic meeting of North and South Korean leaders
- **September 3rd** BBC reports that popular Radio 2 presenter Chris Evans will be leaving the show in December

Acknowledgements

The authors of *Eyes Turned Skyward* have both been involved to one degree or another in restoring aircraft in the collection of the Ulster Aviation Society. But it's been an exercise just as fulfilling and exciting to restore in print and pictures, in however modest a fashion, the adventures and facts of the Society's half-century history. It's been an eye-opening experience and it's provided us both with a new and valuable perspective on our beloved organisation.

We sifted through thousands of pages of reports, newsletters, correspondence, pictures, committee minutes and transcripts seeking facts and photographs and learning to ask the right questions. There were occasional setbacks, but for each one there emerged tales worth telling.

The contributions to this book extend well beyond its two authors, and so we would very much like to thank:
The **Heritage Lottery Fund**, and in particular **Grant Officer Angela Lavin** for providing us with the means and the support needed to complete this project. We now have the foundation for an archive which we hope will be of continuing benefit, not just for the Society, but for aviation historians and enthusiasts throughout all of Ireland and beyond.

Jo-Ann Smyth for turning reams of text and files full of pictures into this beautiful book. Jo-Ann has been incredibly patient with us as we came to grips with the size of the project. The end result is a testament to her skills as a designer.

W&G Baird Printing, and in particular **Gavin Leitch** for printing this book to the highest standard and answering all our awkward questions along the way.

Mark Cairns, UAS design guru, for input, advice and keeping us in check when it came to design decisions.

Ray Burrows, Ernie Cromie, John Barnett, Trevor Haslett, Ross McKenzie, David Hill, Graham Mehaffy, Stephen Boyd, Roger Andrews, Trevor Franklin, John Martin and **Con Law** for taking the time to sit for sometimes lengthy, sometimes winding interviews as we tried to get our heads around the ins and outs of the UAS story. Several of them kindly provided photographs and ephemera as well from their own collections.

Jack Woods for supplying a terrific amount of UAS photographs and memorabilia; they went a long way to getting the project off the ground.

Ernie Cromie for proofreading every line of our lengthy manuscript and keeping the work accurate, fair and balanced at many points along the way.

Fred Jennings, librarian of the UAS, for unearthing the minutes of many meetings, the contents of which proved valuable for our research.

Alan Jarden, Charles McQuillan and **Michael Carbery**, ace photographers, a small sample of whose talent is presented in this volume.

Every **volunteer** of the Ulster Aviation Society, past and present. Many provided occasional suggestions or photographs. Some are named in this book, many more are not, but without every one of them we would not have this wonderful community fifty years on. The support of the Ulster Aviation Society's **management committee**, which endorsed all the projects of our 50th anniversary (including this publication), is especially appreciated. Any errors or omissions in this book are the responsibility of the authors.

Stephen Hegarty would like to thank:
The Fairey Gannet XA460 restoration team for many hours of craic down in Hangar Two and for understanding my commitment to this project over the last year or so.

My parents, **Andrew** and **Evelyn**, whose love and support over the years set me on the path that had me knocking on the Ulster Aviation Society's front door in May, 2013.

My wife, **Kirsty**, for her proofreading of all my contributions to this book, but mostly for her unconditional patience as I gave up evening after evening to get the job done.

And finally, my co-author, **Stephen Riley**, for his input, direction and advice. Stephen's tolerance of my novice writing was a huge relief and thanks to his guidance I've come out the other end of this project a little wiser. It's been a joy to collaborate on bringing the Society's story to its members and beyond.

Stephen Riley would like to thank:
Daughter **Keren Dunbar** for her advice and editing assistance on the long Chapter 5, and daughter **Averil Riley** for her coffee-break kindness and her regular help with our garden, which would be an otherwise innocent victim of my research and writing preoccupations.

My wife, **Mavis**, for her patience and refreshing innocence of all things aeronautical. That said, she is—yes—the wind beneath my fragile wings; just as important, she is the ground beneath my hesitant feet.

My co-author, **Stephen Hegarty**, for his energy, commitment, sense of humour and pesky but necessary reminders about deadlines. Without his positive partnership and genial company, this book would have crashed before take-off.

Picture Credits

CHAPTER	PAGE	IMAGE	CREDIT
Front cover		XT864	Alan Jarden
About the Authors	4	Author portraits	Alan Jarden
Foreword	5	Portrait of Air Vice-Marshal Harvey Smyth	RAF
Frontispiece	7	Alan Moller working on the Fairey Gannet	Charles McQuillan
Introduction	9	Children on a hangar tour	Stephen Riley
Introduction	11	Ariel view of Hangar One	David Doyle
Introduction	12	Aer Turas CL-44 in flight	Fergal Goodman
Introduction	12	UAS in the press	Stephen Riley
Introduction	13	Father and son in the Spitfire	Stephen Riley
Introduction	15	Leonard Craig working on the Phantom	Charles McQuillan
Chapter 1	16	633 Squadron poster	UAS collection
Chapter 1	17	English Electric Lightning in flight	BAE Systems
Chapter 1	17	Collage of magazines	UAS collection
Chapter 1	18	Mannequin in the Royal Observer Corps room	Stephen Riley
Chapter 1	19	David Hill with his Shackleton models	Stephen Riley
Chapter 1	20	Spotters at Belfast City Airport	Stephen Riley
Chapter 1	21	Eddie Franklin climbing down an aircraft ladder	John Barnett
Chapter 1	21	Aircraft recognition poster	UAS collection
Chapter 1	21	Douglas DC-7C at Aldergrove	John Barnett
Chapter 1	22	John Barnett running towards a De Havilland Dove	Eddie Franklin via John Barnett
Chapter 1	23	Air Ulster DC-3 at Aldergrove	John Barnett
Chapter 1	23	Junkers JU-52 at Dublin	Gary Adams
Chapter 1	23	Grumman Goose at Aldergrove	John Barnett
Chapter 1.1	24	Shorts Belfast	Eddie Franklin
Chapter 1.1	24	Aer Lingus 737	UAS collection
Chapter 1.1	24	De Haviland Heron	Jack Woods
Chapter 1.1	24	Philips Blimp	Jack Woods
Chapter 1.1	25	Supper Guppy	Jack Woods
Chapter 1.1	25	Lockheed Jetstar	Jack Woods
Chapter 1.1	25	De Havilland Twin Otter	Jack Woods
Chapter 1.1	25	UAS spotter at Dublin	Jack Woods
Chapter 1.1	25	Ulster Airmail photo pages	UAS collection
Chapter 2	27	Eddie Franklin	via Trevor Franklin
Chapter 2	28	DC-3 in flight	Eddie Franklin via John Barnett
Chapter 2	28	Spotters page from Ulster Airmail	UAS collection
Chapter 2	28	Spotters radio and notebook	Stephen Riley
Chapter 2	28	Security at Aldergrove	Victor Patterson
Chapter 2	29	Halls Hotel bomb damage	via Len Kinley
Chapter 2	30	Avro Shackleton in flight	MoD via John Barnett
Chapter 2	31	Ray Burrows in Corsair wreck	via Raymond Burrows
Chapter 2	32	Corsair wreck in Lough Foyle	Patryk Sadowski
Chapter 2	34	Phantom crashed at Aldergrove	MoD
Chapter 2	34	Woodgate Cherokee	UAS collection
Chapter 2	34	Members view an Aer Turas CL-44	Gary Adams
Chapter 2	35	Antrim Technical College	via Len Kinley
Chapter 2	36	Painting of Wessex over Slemish	Charles McHugh
Chapter 2	38	Mosquito wheel	Ernie Cromie
Chapter 2	38	The UAS tent at Newtownards	Jack Woods
Chapter 2	39	Ernie Cromie	via Ernie Cromie
Chapter 2	40	Lynx on a hilltop	Ernie Cromie
Chapter 2	40	Members carrying wreckage off a mountain	Ernie Cromie
Chapter 2	41	Wreckage of Sunderland DW110	Michael Murtagh
Chapter 2	41	Lockheed Hudson wreckage in the UAS hangar	Stephen Riley
Chapter 2	42	John Hewitt with wreckage	Ernie Cromie
Chapter 2	43	Jack Woods' membership card	Jack Woods

Picture Credits

Chapter 2	43	UAS logo	UAS collection
Chapter 2 .1	44	Irish Aviation Enthusiasts letter	via Des Regan
Chapter 2 .1	44	Gestetner printer	*Grace's Guide to British Industrial History*
Chapter 2 .1	45	*Ulster Airmail #1*	UAS collection
Chapter 2 .1	46	Ray Burrows in UAS tent	Ernie Cromie
Chapter 2 .1	47	*Ulster Airmail* covers	UAS collection
Chapter 2 .1	49	Graham Mehaffy editing *Airmail*	Stephen Riley
Chapter 2 .1	50	2018 *Airmail* cover	UAS collection
Chapter 2 .1	51	*Airmail* being printed	Stephen Riley
Chapter 3	52	Wildcat in Portmore Lough	UAS collection
Chapter 3	53	Wildcat propeller recovery	Ernie Cromie
Chapter 3	53	AGM vote count	UAS collection
Chapter 3	54	Wildcat engine touchdown	Ernie Cromie
Chapter 3	54	UAS constitution	UAS collection
Chapter 3	55	Wildcat airlift	Jack Woods
Chapter 3	55	Wildcat at Ulster Air Show	UAS collection
Chapter 3	57	Painting of Wildcat in flight	David Moore
Chapter 3	58	Wildcat gun camera	Alan Jarden
Chapter 3	59	Peter Lock with Wildcat propeller	Raymond Burrows
Chapter 3	59	Peter Lock's logbook	Peter Lock
Chapter 3	60	Peter Lock wartime portrait	via Peter Lock
Chapter 3	60	Wildcat at Castlereagh Collage	Jack Woods
Chapter 3	61	Members enjoying refreshments	Jack Woods
Chapter 3	62	Trident flightpath	UAS collection
Chapter 3	62	Members boarding the Trident	Jack Woods
Chapter 3	62	DC-3 Poster	UAS collection
Chapter 3	62	Members with DC-3	Jack Woods
Chapter 3	64	Wildcat arriving at the abattoir	Jack Woods
Chapter 3	65	G-BSBH with *Southern Cross*	Jack Woods
Chapter 3	65	Wildcat in Newtownards abattoir	via Raymond Burrows
Chapter 3	66	UAS tent at the Ulster Airshow	Jack Woods
Chapter 3	68	Vampire arrival at Sydenham	Raymond Burrows
Chapter 3	69	Sea Hawk in Shorts Training School	Bombardier
Chapter 3	70	Museum newspaper headline	*Ards Chronicle* via Jack Woods
Chapter 3	70	Sea Fury at Prestwick Air Show	Scott Winning
Chapter 3	71	Sea Hawk being recovered	Raymond Burrows
Chapter 3	72	Stephen Boyd with an Aer Lingus 747	Jack Woods
Chapter 3	74	Control tower tour	Jack Woods
Chapter 3	74	Stephen Boyd with an Aer Lingus Saab 340	Jack Woods
Chapter 3	75	Langford hangar interior	UAS collection
Chapter 3	75	Bedford fuel bowser	Jack Woods
Chapter 3	77	Langford Lodge in wartime	via GI Trail NI
Chapter 3	78	*Ulster Aviation Handbook*	Jack Woods
Chapter 3	79	Shorts SD3-30 with Heyn Van	Jack Woods
Chapter 3	79	View from the DC-8 in flight	Jack Woods
Chapter 3	81	Shorts SD3-30 on the road	Jack Woods
Chapter 3	82	Buccaneer XV361 in service	via Raymond Burrows
Chapter 3	83	Buccaneer XV361 delivery flight	UAS collection
Chapter 3	83	Buccaneer XV361 at Langford	UAS collection
Chapter 3	84	Wessex at Langford Open Day	Jack Woods
Chapter 3	85	1995 Open Day leaflet	UAS collection
Chapter 3	85	1995 Open Day leaflet interior	UAS collection
Chapter 3	87	B17 over Gartree church	Jack Woods
Chapter 3	88	Hot air balloon at Langford Lodge	Jack Woods
Chapter 3	89	Air Force One at Aldergrove	Jack Woods
Chapter 3	89	USS JFK In Dublin Bay	Jonathan McDonnell

Picture Credits

Chapter 3	90	UAS group with the Vickers Merchantman	Jack Woods
Chapter 3	90	Vickers Merchantman video cover	UAS collection
Chapter 3	91	UAS tickets and advertisements	via Jack Woods & Colin Boyd
Chapter 3.1	93	Team working on the Wildcat	Ernie Cromie
Chapter 3.1	94	Ernie Cromie on the Wildcat wing	via Ernie Cromie
Chapter 3.1	94	Wildcat on the shore	Jack Woods
Chapter 3.1	95	Wildcat airlift	Jack Woods
Chapter 3.1	96	Team preparing the SD3-30	Jack Woods
Chapter 3.1	97	SD3-30 airlift	Jack Woods
Chapter 3.1	98	Phantom wings being loaded	Stephen Riley
Chapter 3.1	99	Phantom in the Leuchars hangar	Stephen Hegarty
Chapter 3.1	101	Phantom wings on the road	Stephen Hegarty
Chapter 4	102	Stephen Boyd and Jack Woods	via Jack Woods
Chapter 4	103	Fred Jennings in the library	Stephen Riley
Chapter 4	103	Book covers	UAS collection
Chapter 4	104	Guy Warner books signing	Stephen Riley
Chapter 4	104	Antonov AN-124	Jack Woods
Chapter 4	105	EasyJet 737	Aero Icarus
Chapter 4	106	UAS trip to Baldonnel	Jack Woods
Chapter 4	107	Iolar	UAS collection
Chapter 4	107	Paddy Crowther	Jack Woods
Chapter 4	108	Ernie Cromie with John Hutchinson	UAS collection
Chapter 4	108	UAS website header	UAS collection
Chapter 4	109	UAS members with helicopter crews	Jack Woods
Chapter 4	109	UAS signpost	Jack Woods
Chapter 4	111	Langford Lodge hangars	Dorothy Bunting
Chapter 4	112	Visitors reading Wildcat displays	Jack Woods
Chapter 4	112	Irish Air Corps Alouette	Jack Woods
Chapter 4	113	UAS leaflet	via Jack Woods
Chapter 4	113	Piper Cub at Langford	UAS collection
Chapter 4	114	Grumman Albatross at Langford	UAS collection
Chapter 4	114	Tucano Christmas card	via Jack Woods
Chapter 4	115	Gartree Church	Jack Woods
Chapter 4	115	Sir Philip Foreman	UAS collection
Chapter 4	116	Ernie Cromie crushing cans	Jack Woods
Chapter 4	116	Buccaneer outside Langford hangars	UAS collection
Chapter 4	117	Irish Air Corps Dauphin	UAS collection
Chapter 4	117	Peter Lock with the Wildcat	UAS collection
Chapter 4	119	Langford control tower	Dorothy Bunting
Chapter 4	120	Vampire in Langford Lodge	UAS collection
Chapter 4	120	Spitfire at Landford Lodge	UAS collection
Chapter 4	121	Langford display room	UAS collection
Chapter 4	123	G-BTUC in flight	Jack Woods
Chapter 4	124	G-RENT in Langford	UAS collection
Chapter 4	124	Eurowing Goldwing in the Maze/Long Kesh	UAS collection
Chapter 4	125	Jet Provost arrives at Langford	UAS collection
Chapter 4	125	Langford model collection	UAS collection
Chapter 4	126	Wessex XR517 in service	via Raymond Burrows
Chapter 4	126	Wessex Fund display	Jack Woods
Chapter 4	127	Ray Burrows with Wessex	UAS collection
Chapter 4	128	Marbur Properties eviction letter	via Ernie Cromie
Chapter 4	129	John Miller on stage	UAS collection
Chapter 4	130	Warren Bradley with the mayor and re-enactors	Dorothy Bunting
Chapter 4	132	Map of potential museum sites	Open Street Maps
Chapter 4	133	Long Kesh in 1942	UAS collection
Chapter 4	134	Vampire at Clotworthy	Dorothy Bunting
Chapter 4	135	Turret restoration in progress	Eric Gray & UAS collection

Picture Credits

Chapter 4	136	Map of the Maze/Long Kesh site	UAS collection
Chapter 4	136	Sea Hawk tow	Eric Gray
Chapter 4	137	First aircraft inside the Long Kesh hangars	UAS collection
Chapter 4	137	Broken windows at Long Kesh	UAS collection
Chapter 4	138	SD3-30 wing removal	Neville Greenlee
Chapter 4	139	Buccaneer lift	Eric Gray
Chapter 4	139	SD3-30 & Buccaneer in transit	Eric Gray
Chapter 4	141	Buccaneer arrival at Long Kesh	Neville Greenlee
Chapter 4.1	143	Scout being unloaded	Stephen Riley
Chapter 4.1	145	Ferguson assembly	Stephen Riley
Chapter 4.1	147	Crowds at Portrush 2017	Stephen Riley
Chapter 4.1	148	Portrush shop	Stephen Riley
Chapter 4.1	149	William McMinn and Ferguson	Stephen Riley
Chapter 5	150	UAS hangars from the air	UAS collection
Chapter 5	151	Maze/Long Kesh development plan	UAS collection
Chapter 5	151	Member looking through a broken window	UAS collection
Chapter 5	152	Hangar room interior	UAS collection
Chapter 5	152	McKillop Award programme	via Des Regan
Chapter 5	153	Paul McMaster with the DC-3	UAS collection
Chapter 5	153	*Battle for the Skies* DVD cover	UAS collection
Chapter 5	154	SB.4 Sherpa arriving at the Maze/Long Kesh	UAS collection
Chapter 5	155	Jet engine turbine blades	Alan Jarden
Chapter 5	157	Ferguson flight painting	Gerald Law
Chapter 5	158	Playing cards	Alan Jarden
Chapter 5	158	Vampire, dismantled	UAS collection
Chapter 5	158	Harry Munn and Jim Robinson	Stephen Riley
Chapter 5	159	Hangar Christmas party	Stephen Riley
Chapter 5	159	Harry Ferguson float	UAS collection
Chapter 5	161	XH131 in flight	Bombardier
Chapter 5	162	David Jackson on XH131	Mal Deeley
Chapter 5	162	XH131 wing lift	Mal Deeley
Chapter 5	163	XH131 arrival at the hangars	Mal Deeley
Chapter 5	163	Ray, Mal and David with XH131	via Mal Deeley
Chapter 5	164	Coast Guard helicopter at an open day	UAS collection
Chapter 5	164	Lilian Bland	UAS collection
Chapter 5	165	Ray Burrows presents Ernie Cromie with a painting	Stephen Riley
Chapter 5	166	Ray Burrows cleaning the Canberra	Stephen Riley
Chapter 5	166	Wildcat undercarriage indicators	Stephen Riley
Chapter 5	167	Radial engine crank shaft	Alan Jarden
Chapter 5	167	Sherpa restoration team	Stephen Riley
Chapter 5	168	Ray Burrows briefing hangar volunteers	Stephen Riley
Chapter 5	168	Open day whiteboard plan	Stephen Riley
Chapter 5	169	Ernie vacuuming	Stephen Riley
Chapter 5	169	Mannequin	Charles McQuillan
Chapter 5	171	Crowd in the hangar on open day 2012	UAS collection
Chapter 5	172	The public view the Wildcat	Stephen Riley
Chapter 5	173	The Buccaneer at open day 2013	Stephen Riley
Chapter 5	173	Film crew at open day 2013	Stephen Riley
Chapter 5	174	Peter Robinson and Martin McGuinness	Stephen Riley
Chapter 5	175	Peace Centre billboard	via Stephen Riley
Chapter 5	175	Walter McManus	via Stephen Riley
Chapter 5	176	Spitfire manufacture	via Stephen Riley
Chapter 5	176	Spitfire arrival at UAS hangars	Stephen Riley
Chapter 5	177	Roy Kerr with the SD3-30 propeller	Stephen Riley
Chapter 5	177	Flying helmet	Alan Jarden
Chapter 5	178	Thornycroft Amazon	Alan Jarden
Chapter 5	178	Seacat missile	Alan Jarden

Picture Credits

Chapter 5	179	Ernie Cromie BEM ceremony	via Ernie Cromie
Chapter 5	179	Spitfire assembly trial	Stephen Riley
Chapter 5	179	Spitfire in the news	UAS collection
Chapter 5	181	Spitfire unveiling	Stephen Riley
Chapter 5	182	Puma arrival	Stephen Riley
Chapter 5	182	Phantom XT864 at Leuchars	Mal Deeley
Chapter 5	183	The Gannet crew	Stephen Riley
Chapter 5	183	V/1500 propeller hub	Alan Jarden
Chapter 5	183	David Jackson	Stephen Riley
Chapter 5	184	Argus recovery	Stephen Riley
Chapter 5	184	William and Nathaniel Smyth	Stephen Riley
Chapter 5	185	Ron Bishop with a tour group	Stephen Riley
Chapter 5	185	Children in the Alouette	Stephen Riley
Chapter 5	186	Harry Ferguson	UAS collection
Chapter 5	187	Stephen Lowry and William McMinn	Stephen Riley
Chapter 5	187	BBC Ferguson handover	Press Eye via the BBC
Chapter 5	189	Ferguson Flyer in the air	Charles McQuillan
Chapter 5	190	First aid training	Stephen Riley
Chapter 5	191	ATC Sting assembly	Stephen Riley
Chapter 5	192	Belfast Metropolitan Collage students	Stephen Riley
Chapter 5	192	Film crew in the hangar	Stephen Riley
Chapter 5	193	Belfast Film Festival *Airplane!* screening	Alan Jarden
Chapter 5	193	Technical manual	Alan Jarden
Chapter 5	195	DeLoreans visiting the UAS hangars	Stephen Riley
Chapter 5	196	Paul Young at GI Jive	Stephen Riley
Chapter 5	197	Fionnuala Fagan	Stephen Riley
Chapter 5	198	Billy McCall and Harvey Smyth	Stephen Riley
Chapter 5	198	Technical drawing	Stephen Riley
Chapter 5	199	Sea Prince recovery	Raymond Burrows
Chapter 5	199	Ian Hendry and the Wildcat	Stephen Riley
Chapter 5	200	Tucano team	Stephen Riley
Chapter 5	201	Argus Team	Stephen Riley
Chapter 5	201	How planes fly	Stephen Riley
Chapter 5	202	Peter Morrison and the Sherpa	Stephen Riley
Chapter 5	203	UAS members at Le Bourget	Alan Jarden
Chapter 5	203	Mirage display at Le Bourget	Alan Jarden
Chapter 5	205	Phantom unveiling	Alan Jarden
Chapter 5	206	Meteor tail section	Stephen Riley
Chapter 5	207	Portrush 2018	Alan Jarden
Chapter 5	207	Argus, painted	Stephen Hegarty
Chapter 5	209	UAS members with the Lord Lieutenant	Michael Carbery
Chapter 5	208	Queen's Award crystal	Michael Carbery
Chapter 5.1	210	V/1500 propeller	Darren Sprules
Chapter 5.1	211	The Ernie Cromie Room	Alan Jarden
Chapter 5.1	211	The World War One Room	Alan Jarden
Chapter 5.1	212	Paddy Malone	Stephen Riley
Chapter 5.1	213	The Sir Philip Foreman Room	Alan Jarden
Chapter 5.1	213	The Fred Jennings Radio Room	Alan Jarden
Chapter 5.1	215	Hangar map	Gráinne Hegarty
Chapter 5.1	216	The Aldergrove Room	Alan Jarden
Chapter 5.1	217	The Engine Room	Alan Jarden
Chapter 5.1	218	Shorts SD3-30 model	Alan Jarden
Chapter 5.1	219	23MU reunion	Alan Jarden
Chapter 5.1	219	Wildcat display	Alan Jarden
Chapter 6	220	Eddie Franklin at Sheppey	via Trevor Franklin
Chapter 6	221	Belfast spotters meet	Stephen Riley
Chapter 6	223	Museum concept sketch	William A.J. Dawson
Chapter 6	224	UAS members with a Tornado	UAS collection
Chapter 6	224	UAS members painting	Stephen Riley
Chapter 6	227	Hangar future concept sketch	Gerald Law
Back cover		Wildcat	UAS collection